T0199892

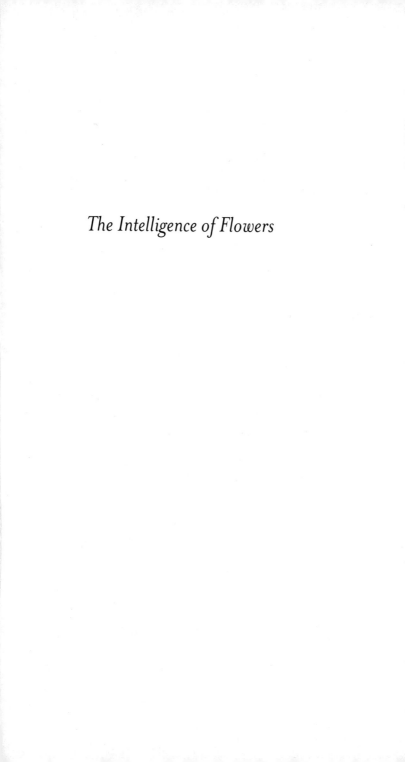

The Intelligence of Flowers

The Intelligence of Flowers

MAURICE MAETERLINCK

*Translated and
with an Introduction by
Philip Mosley*

STATE UNIVERSITY OF NEW YORK PRESS

COVER PHOTOGRAPH: *Orchid No. 1*, 2005, ANDREW SOVJANI
used by permission

Published by
STATE UNIVERSITY OF NEW YORK PRESS
Albany

For information, contact
STATE UNIVERSITY OF NEW YORK PRESS
www.sunypress.edu

Production and book design, Laurie Searl
Marketing, Susan M. Petrie

Library of Congress Cataloging-in-Publication Data
Maeterlinck, Maurice, 1862–1949.
[Intelligence des fleurs. English]
The intelligence of flowers / Maurice Maeterlinck ;
translated and introduction by Philip Mosley.
p. cm.
Includes bibliographical references.
ISBN 978-0-7914-7273-6 (hardcover : alk. paper)
ISBN 978-0-7914-7274-3 (pbk. : alk. paper)
1. Botany. I. Mosley, Philip. II. Title.

PQ2625.A4I513 2008
844'.8–DC22
2007001782

10 9 8 7 6 5 4 3 2 1

Contents

Acknowledgments

Initial work on the translation was facilitated by a residency at the European College of Literary Translators in Seneffe, Belgium. Thanks are due to Françoise Wuilmart, director of the college, and to Jean-Luc Outers, Senior Literary Counselor in the Belgian Ministry of the French Community, Brussels. Support for travel to this residency came from the office of Academic Affairs at the Worthington Scranton campus of Penn State University, and from the Office of International Programs at Penn State. Thanks also to Frans De Haes at the Archives and Museum of Literature, Brussels, for his assistance. I am grateful to Elinor Shaffer and Ashton Nichols for their helpful comments on an early draft of the introduction. Finally, thanks to Shu-ching, my wife, who helped in many ways.

Introduction

Relatively neglected since the mid-twentieth century, the Belgian author Maurice Maeterlinck (1862–1949), winner of the Nobel Prize for Literature in 1911, was once one of the most widely read authors in the world. He was appreciated particularly in Britain, the United States, and Germany. Though a writer in French of Flemish origin, he remained rather less influenced by French literature and culture than by those of Germany and Britain, which conformed more closely to his Flemish sensibility and Germanic cast of mind. After the end of World War I he stayed briefly in the limelight, notably touring the United States and spending time in Hollywood. Yet his best work was already well behind him. Another brief spell of public attention accompanied his return to the United States, where he lived from 1940 to 1947, but after his death in 1949 he became for the most part a forgotten figure.

We remember Maeterlinck today as a symbolist pioneer of modern drama and as a poet. His extraordinary literary success came firstly from the innovative and strangely atmospheric plays he wrote in the last decade of the nineteenth century, among them *Princess Maleine*

(1889), *The Intruder* (1890), and *The Blind* (1890). *Pelleas and Melisande* (1892) inspired a famous opera by Claude Debussy (1902) and was also set to music by Gabriel Fauré, Jean Sibelius, and Arnold Schoenberg. In the next phase of his career, beginning with the twentieth century, Maeterlinck showed major changes in his artistic interests and personal outlook, while bolstering his dramatic reputation with *The Blue Bird* (1909), his most enduring theatrical accomplishment.

Maeterlinck's plays won him international acclaim, and he proceeded to cement his fame with *The Treasure of the Humble* (1896) and *Wisdom and Destiny* (1898), two collections of philosophical essays, and a hugely successful book-length essay, *The Life of the Bee* (1901). Maeterlinck proved to be an accomplished essayist, and it was this part of his literary output, rather than his extraordinary plays, that largely maintained his global reputation, fixing him in the hearts and minds of most of his readers to the greatest extent over time. *The Life of the Bee*, his first extended study of the workings of the natural world, sold over a quarter of a million copies, including translations into many languages. It led to three other studies of its kind: the eponymous essay in *The Intelligence of Flowers* (1907) that forms the main part of the present translation; *The Life of the Termite* (1926), which sold eighty thousand copies in two years; and *The Life of the Ant* (1930), of which seventy thousand copies were printed by its publisher. "The Intelligence of Flowers" plus a related essay on scents, also translated here, comprise about one-third of an original volume of eleven essays embracing various topics from boxing to immortality. In this respect *The Intelligence of Flowers* differs from the other three works on nature, each of which is an entire volume on its subject divided into subtitled sections.

Given that Maeterlinck's essays assured him of a vast readership long after his plays had all but faded from public view, it is revealing of current critical judgment that he is not included in a recent one thousand-page encyclopedia of the essay as literary genre. This is all the more surprising, since Maeterlinck owed his Nobel Prize mainly to his essay collections. The Swedish Academy had twice refused Maeterlinck the prize on account of its resistance to the symbolism and fatalism of his early work, but in 1911 it finally recognized his exceptional impact on modern literature.

Maeterlinck's mastery of the essay form corresponded to a great increase in the popularity of this genre in French in the late nineteenth and early twentieth centuries. Speculative in the nineteenth-century tradition and imbued with an elegant and idiosyncratic touch, Maeterlinck's free-flowing style of personal expression was ideal for his time. By understanding the essay form as an ideal means of conveying philosophical ideas and specialized knowledge in an appealing way, Maeterlinck conceived "The Intelligence of Flowers" as an intellectual vulgate presented with the craft and sensitivity of a poet. Balancing the scientific and artistic, emotional and cerebral, specific and general, his essay is a botanical tour de force written in lyrical and accessible prose.

Before introducing "The Intelligence of Flowers" in more detail, it may be useful to consider briefly the debates on the relationships of literature to science in general and of literature to botany in particular.

Work within the humanities since the 1960s, much of it under the sway of structuralist and poststructuralist theories, has brought fresh insights into the ways in which the sciences and the humanities may interact. Many scholars argue that the gap between scientists and nonscientists has never been quite as wide as has often been suggested. We may see the questioning of objectivity in quantum physics by scientists, for instance, as concurrent with the questioning of the nature of scientific enquiry by philosophers, historians, literary critics, and art critics in the light of new propositions about meaning, discourse, and textuality. These scholars look askance at a highly polarized model, recognizing instead a continuing exchange and encounter in the movements across artistic and scientific fields as well as in the construction of their respective discourses.

The emergence of several new critical methods may help to regenerate interest in Maeterlinck's work on nature. Stimulated by the liberal tenor of much poststructuralist theory, ecocriticism (or "green" criticism) is an interdisciplinary methodology that has burgeoned since the late 1980s. Broadly concerned with interconnections between nature and culture, ecocriticism has spawned in turn a literary subdivision known as ecoromanticism, which studies Romantic writers' views of the natural world. Ecoromanticism is relevant to the study of Maeterlinck's nature writing in view of the great influence on him of Romantic philosophy. An alternative view that challenges relativistic approaches to science and the humanities has come from sociobiologists, who believe that art has a biological basis. A literary subdivision of this school of thought, known as biopoetics (or Darwinian criticism), shares this belief.

For the literary scholar seeking further evidence of interdisciplinary fruitfulness there are several other good starting points: the evolution of science fiction, or the reconnection of literature and medicine within the medical humanities, or the successful communication of scientific knowledge in sophisticated journalism and in mass-market books by figures such as Edward O. Wilson, Stephen Hawking, and Stephen Jay Gould. This popular dissemination of science—a phenomenon referred to as the Third Culture—represents in many respects the survival of a nineteenth-century tradition of popularized knowledge of a kind that Maeterlinck recast in his own essays on insects and plants.

If we trace botanical science to its origins in ancient Greek philosophical discourse, we find that Theophrastus and Dioscorides established the first taxonomies, while Plato attributed desires to plants in a foreshadowing of Arthur Schopenhauer's theory of will in nature. In Roman times Virgil and Lucretius were among the earliest literary figures to utilize the imagery of plants and flowers in their works. Botany stagnated until sixteenth-century herbalists, often considered the fathers of modern botany, began to seek better methods of plant identification. Their quest culminated in Carolus Linnaeus's revolutionary eighteenth-century system of classification, a milestone that ushered in an age dominated by systematics (scientific nomenclature and classification). The eighteenth-century materialist Julien Offray de La Mettrie described many common elements of human and plant life, while Erasmus Darwin (grandfather of Charles), following Linnaeus, speculated on the emotional and sexual life of plants.

However, botany still retained its basic forms until the birth of modern biology in the early nineteenth century when Lamarckian evolutionism, followed closely by

Darwinism, revolutionized the entire study of the natural world. As Bertrand Russell puts it, "The prestige of biology caused men whose thinking was influenced by science to apply biological rather than mechanistic categories to the world. Everything was supposed to be evolving, and it was easy to imagine an immanent goal."* This vision appealed strongly to Romantic philosophers and poets. Modern botany was thus shaped in part by a philosophical idealism rooted in organicist and pantheistic ideas. Such ideas were themselves remnants of premodern science, which, as Michel Foucault and others have argued, did not privilege observation over other kinds of explanations of the world. Nature philosophy, which emerged primarily from German Romanticism, placed great store on the idea of a universal force or spirit animating all living things. Such an antimechanistic idea was closely identified with that of a world soul, a concept traceable to Plato, Pythagoras, and the Stoics. Applied to botany, nature philosophy produced, for instance, *Nanna, or the Soul-Life of Plants* (1848), in which Gustav Theodor Fechner argued for both physical and psychical life in plants.

Jean-Jacques Rousseau and Jacques-Henri Bernardin de Saint-Pierre initiated a new literary discourse of botany, one from which Maeterlinck would adopt the idea of nature and art as created and driven by a common force. In *The Reveries of a Solitary Walker* (1782), in *Letters on Botany* (composed 1771–1773), and in several other botanical reference works, Rousseau unfolded a proto-Romantic vision of unspoiled nature wherein the

---

*Bertrand Russell, *History of Western Philosophy* (London: George Allen & Unwin, 1962) 698.

lone roamer reflects upon his place in the greater scheme of things. The equally lyrical style of Bernardin's *Studies of Nature* (1784), notable for its classification of trees, also proved influential on the composition of "The Intelligence of Flowers," though both Bernardin's utilitarian program and his deistic belief would fail to impress the agnostically minded Maeterlinck, who refers to them in chapter three of his essay as "naive errors" to be avoided.

The literary-botanical discourse was developed more systematically by Johann Wolfgang von Goethe and John Ruskin, though Ruskin carefully distanced himself from its more mystical aspects. Goethe, whose botanical studies began in 1775 shortly after his move to Weimar, became the key figure in the development of a philosophical-scientific study of plant life, and it was he who coined the term *morphology*. In works such as *The Metamorphosis of Plants* (1790), *Studies for a Physiology of Plants* (c. 1795) based on discussions with Friedrich Schiller, and especially *Italian Journey* (undertaken 1786–1788, published 1816–1829), Goethe searched among plants for an archetype (Urpflanze) and for evidence (following Baruch Spinoza) of a divine force at work in nature. Writing to Johann Gottfried von Herder from Naples in May 1787, he spoke excitedly of being "very close to the secret of the reproduction and organization of plants," to proof of their "inner necessity and truth" according to a law that "will be applicable to all other living organisms."* Herder had already established a botanical model for aesthetics as early as 1778 in *On the Knowing and Feeling of the*

*Johann Wolfgang von Goethe, *Italian Journey* (1786–1788), translated by W. H. Auden and Elizabeth Mayer (San Francisco: North Point Press, 1982) 305–6.

Human Soul. Working later within the same panpsychical framework, Samuel Taylor Coleridge based a large part of his *Theory of Life* (1848) around plant analogies, while Thomas Carlyle thought of plant growth as comparable to artistic creation.

Ruskin's *Proserpina: Studies of Wayside Flowers* (1874) collects work begun as early as 1842 and further developed in 1868. In a witty and idiosyncratic metatext invoking art, literature, philosophy, and religion, he sought to make botany more enjoyable than in the context of orthodox science with its complex system of nomenclature. For Ruskin, botany was as much biographical as descriptive. In the Preface to *Modern Painters* (1844), he claims "the difference between the mere botanist's knowledge of plants, and the great poet's or painter's knowledge of them" is that "the one notes their distinctions for the sake of swelling his herbarium, the other, that he may render them vehicles of expression and emotion."[*]

Let us also remember that from the late eighteenth century a visual art modeled on plants developed alongside this literary discourse. Botanical illustrators explored the aesthetic potential of plant life in paintings, drawings, and engravings, later exercising a decisive influence on the scientific romanticism of art nouveau and on the plant designs William Morris contributed to the aesthetics of the arts and crafts movement. Among famous art nouveau designers, Emile Gallé was also a botanist, while Eugène Grasset (in *Plants and Their Application to Ornament*, 1897) and Ernst Haeckel (who coined *ecology*) were hugely influential in defining

---

[*]Cited by M. H. Abrams, *The Mirror and the Lamp: Romantic Theory and the Critical Tradition* (New York: Norton, 1958) 312.

formal variations of the style. Maeterlinck thus grew up in a blooming botanical culture. The Royal Botanical Society of Belgium had been founded in 1862, the year of his birth, and in 1870 the Brussels Botanical Garden was opened. For the young Maeterlinck botany was a surging science that also promised untold artistic delight.

Maeterlinck's four principal essays on nature are among the last examples of a certain kind of humanistic science to be appreciated in Western culture up to the early twentieth century. In each of his three entomological studies Maeterlinck reconstructs a model first shaped by ancient Greek philosophy: an intelligent natural society comparable in many respects to that of mankind. In "The Intelligence of Flowers" he draws similar analogies between the organization and inventiveness of the floral and human worlds. The discourse of Maeterlinck's nature essays, based on philosophical and artistic responses to the methodical observation of nature, gradually fell out of fashion as the dialogue between science and the humanities—despite spirited defenses by Albert Einstein, Alfred North Whitehead, Aldous Huxley, and others—gave way to new specialist and technocratic norms.

"The Intelligence of Flowers" belongs to a period of about fifteen years in Maeterlinck's career from 1897 onward. Emerging from a pessimism that nonetheless had engendered his major symbolist plays, Maeterlinck entered a transitional phase still inspired by Schopenhauer but in more positive, optimistic, and cheerful ways. This period became clear in Maeterlinck's preface to Paul Lacomblez's 1901 edition of his plays, a statement that announced a curious and encouraging blend of science, spiritualism, and occultism henceforth to be his preoccupation. Maeterlinck's Nobel Prize in 1911 marked the height of his fame after which he returned to a mood of fatalistic doubt, confirmed by his 1911 essay collection *Death* (published in 1913). By the time he published his essays on the termite and the ant, his view of these natural societies suggests failure and despair rather than perfect harmony. The termite world is no longer a role model for mankind, but closer to the dystopian visions later fictionalized by Aldous Huxley and George Orwell.

During his brief sunny period a botanical passion helped to instill in him a certain humility and contentment, and he spent much time after 1903 in his house Les Quatre Chemins (The Four Ways) near Grasse in southern France, where he pursued his study of flowers, especially those produced locally for perfume, and carried out experiments in hybridization. Some early reflections found their way, amid other subjects, into *The Double Garden* (1904), but it was in "The Intelligence of Flowers" that Maeterlinck expressed his botanical ideas in full.

"The Intelligence of Flowers" is a detailed record of a philosophical naturalist at work, and it follows *The Life of the Bee* in showing Maeterlinck, a famous man of letters, very much an inheritor of the Romantic notion of poet as medium, seeking to present (or popularize, if you will) scientific knowledge to his large audience. That this enterprise came wrapped in a cosmological package revealed both his temperament and the eclecticism of much turn-of-the-century culture. His contemporary Edward Thomas (1911) calls him a "mystic man of the world," while present-day scholar Patrick McGuinness chooses the term "boudoir mystic" to describe

Maeterlinck's "vague agnostic spirituality."* The essay on the intelligence of flowers integrates several discourses: science, literature, philosophy, and (rather paradoxically) an earthy mysticism, all skillfully sutured by Maeterlinck's rhetorical command and fluent style. Even though Thomas retains doubts about the author's "starry urbanity" and "science in fancy dress," he acknowledges that Maeterlinck's descriptions "can be masterly, more brief and precise at their best than those of Ruskin, with which alone they can be compared."†

Deriving many ideas from Schopenhauer's *On the Will in Nature* (1836), Maeterlinck asserts that flowers possess thought without knowledge, a capacity that nonetheless, as in bees, constitutes a form of intelligence. Maeterlinck associates this belief in vegetable intelligence with (to invoke William Wordsworth's "Lines Written a Few Miles above Tintern Abbey," 1798) his own "sense sublime / Of something far more deeply interfused / Whose dwelling is the light of setting suns," a sense in accord with his view of human knowledge as the highest but by no means exclusive agent of a supreme order, of "a motion and a spirit, that . . . rolls through all things" (Wordsworth again).‡ Maeterlinck was directly influenced both by the German Romantics (particularly Novalis, whom he translated into French) and by certain

---

*Patrick McGuinness, *Maurice Maeterlinck and the Making of Modern Theatre* (Oxford: Oxford University Press, 2000) 6.

†Edward Thomas, *Maurice Maeterlinck* (New York: Dodd, Mead, 1911) 255, 267.

‡See Wordsworth and Coleridge, *Lyrical Ballads 1798*, edited by W. J. B. Owen (Oxford: Oxford University Press, 1969) 114–15.

of their exponents in the English-speaking world, notably Thomas Carlyle and Ralph Waldo Emerson.

Maeterlinck's corresponding belief in all life forms as immortal and in that which creates them as being intellectually unknowable reveals the influence on him also of mystical and gnostic thought: Jakob Böhme and Emanuel Swedenborg, for instance, and especially his fellow Fleming, Jan van Ruusbroec. Maeterlinck's love of the fourteenth-century mystic's work (he translated *The Adornment of Spiritual Marriage* into French in 1891) also bolstered an identification with the Germanic mind that he had begun to feel several years before Octave Mirbeau's 1890 review of *Princess Maleine* in *Le Figaro* catapulted him to sudden fame.

Yet, practical, reasonable man of the world that he was, Maeterlinck would only go so far down a mystical road. He showed little inclination to true mystical detachment from the material world. Despite his Romantic sensibility, he was by no means opposed to reason or science. Maeterlinck's belief in the possibility of a mystical reason suggests the intuitive positivism of Henri Bergson, who was perhaps the closest of contemporary philosophers to Maeterlinck's way of thinking. Fully aware that the discoveries of modern biology exposed the limits of neoplatonism, Maeterlinck took an opportunity in "The Intelligence of Flowers" to incorporate elements of Darwinism into his holistic vision. In doing so he followed a series of works (*The Fertilization of Orchids*, 1862; *Variation in Animals and Plants under Domestication*, 1868; *The Power of Movement in Plants*, 1880) in which Darwin grants to plants near-human powers in their ability to fight for survival, to reproduce, and to vary in form. Though many of Maeterlinck's botanical references are from traditional sources, he cites, in addition to Darwin, the experiments of two contemporary French botanists, Gaston Bonnier and E-L. Bouvier. Given the extent of Darwin's influence by the turn of the century, it would seem strange if Maeterlinck had not followed him in many of his scientific ideas (as did H. G. Wells, for instance, in his 1895 story, "The Flowering of the Strange Orchid"). After all, Darwin himself was greatly influenced by social philosophy and by literature, and it is unsurprising to learn how much Darwin's own botanical work owed to ideas he drew from creative writers, and in particular, like Ruskin, from the poetry of John Milton.

It is tempting to suggest that had Maeterlinck been less inclined early in his career to practice suggestive and intuitive thinking, to relish hidden correspondences and esoteric visions—all that made him, like the young William Butler Yeats, an exemplary symbolist writer—he might instead have found himself at home in naturalism and, like Émile Zola, applying strict scientific methods to his literary art. Present-day critic Paul Gorceix resolves this apparent duality by insisting on a methodological unity between the symbolist poet-dramatist and the scientific naturalist in Maeterlinck, thus placing his work on nature in an overarching field: the history of ideas. In Gorceix's view, there is no creative break between these two identities. Maeterlinck's close observation of plants led him to speculate analogically on the mysteries of life in man and the universe, and his work "constitutes a whole, of which the cement is the epistemology of totality."[*]

---

*Paul Gorceix, "Postface," in Maurice Maeterlinck, *La Vie de la nature* (Brussels: Editions Complexe, 1997) 501. All translations from French are my own.

As for the question of evolution, Maeterlinck's ability to reconcile art, science, and philosophy in his literary method allows us to place him between the opposite poles of scientific positivism and religious belief. In section XXIII of "The Intelligence of Flowers," he attempts a definition of evolution: "Does not this rather vague word mean, in the final analysis, adaptation, modification, intelligent progress?" For Maeterlinck these processes are entirely consistent with a Romantic view of the universe, for they are driven in flower and man alike by a common spirit: "We follow the same path as the soul of this great world" (section XXIX). And though he speaks almost in the next breath of a universe "molded by unknown substances," he attempts, in the final section of his essay, to specify this all-encompassing spiritual force in a formulation that brings him as close to religious belief as he allows: "A scattered, general intelligence, a kind of universal fluid that penetrates diversely the organisms it encounters, depending on whether they are good or bad conductors of consciousness. Mankind would represent, until now, upon this earth, the realm of life that offered the least resistance to this fluid that the religions call divine."

Even if we choose not to accept Maeterlinck's curious blend of science and spiritualism, he occupies nonetheless a special place in the progressive literary and intellectual genealogy of the early twentieth century. We are offered in a highly palatable form the findings of an unsystematic, rather old-fashioned thinker, endowed with an exceptional literary talent for enlivening the study of botany and for making it not only a more accessible science but also, rather quaintly from our contemporary viewpoint, part of a quest for all-encompassing but ultimately unverifiable truths. This position seems nonetheless at one with the respect accorded in the late nineteenth century to a generalist spirit of knowledge in which little conflict occurred between the claims of science and the capacities of the imagination. This coexistence of seemingly opposed beliefs bears witness to the presence of an episteme (to use Foucault's term), a discursive space specific to the period, one in which appropriate conditions existed for the construction, articulation, and reception of various currents of knowledge and thought. In the late intellectual life of Victorian Britain, for instance, literary figures such as Samuel Butler and Frederic W. H. Myers, as well as a heterodox group of scientific theorists, practiced a philosophy of honest doubt, rejecting institutional religion and scientific naturalism but not the potential union of science, spirit, and the imaginative life.

"The Intelligence of Flowers" impressed some modernist authors. Marcel Proust frequently cited Maeterlinck's work, and he drew directly on it for his analogy of orchid reproduction in the first part of "Sodom and Gomorrah," the fourth volume of *In Search of Lost Time*. Most early critics of "The Intelligence of Flowers" were also impressed by Maeterlinck's expression of more rational and ethical ideas as well as by his bravura performances as a poet of science. B. Timmermans (1912), for instance, suggests that Maeterlinck's nature essays indicate a moral rather than an intellectual quest, and thus to judge him as a dilettante is to ignore his belief in the moral necessity of intelligent work irrespective of scientific results or of limits to human knowledge. Yet many of these early critics also share misgivings about Maeterlinck's scholarly attributes. Among them are a few vehement opponents for whom neither moral incentive nor poetic talent may compensate for scientific and philosophical weaknesses. Maurice

Lecat (1937), for example, derides Maeterlinck's poor documentation (or rather, lack of it), wheeling up two big guns, Massart and Bonnier, to bolster his attack. Ironically enough, Bonnier, to whose experiments on toxin secretion Maeterlinck refers in "The Intelligence of Flowers," had earlier been generous in his praise of *The Life of the Bee*.

A similarly mixed response persists in later criticism of Maeterlinck's essay. Robert Vivier (1964) reiterates the view that Maeterlinck acknowledged a need to stay within practical human bounds, an opinion shared by Belgian modernist author Franz Hellens based on his own conversations with Maeterlinck. For Vivier, Maeterlinck seeks "wisdom and happiness, and occasionally at the most a quarter-hour of light dizziness on the balcony of hypothesis."[*] His careful observation of the natural world serves to check the possible excesses of his transcendent imagination. W. D. Halls (1960) reinforces this opinion: the decline of Christianity prompts Maeterlinck to seek a new moral code, a search that sets the tone of "The Intelligence of Flowers" and leads its author toward a humanist and, by 1913, an overtly socialist position.

Halls points to Maeterlinck's superficial knowledge of Darwinism, claiming that the author is rather less successful as a botanist than an entomologist. Indeed, while Maeterlinck's career coincided with a number of important developments in biology, it appears unlikely that he was fully aware of those in microbiology, evolutionary taxonomy, biochemistry, and morphology and, most importantly, following the 1900 rediscovery of Gregor Mendel's 1866 paper, of those leading to the birth of genetics.

---

[*]Robert Vivier, "Deux aspects de Maeterlinck," in *Le Centenaire de Maurice Maeterlinck* (Brussels: Palais des Académies, 1964) 235.

Within the field of botany, the American horticul-
turalist Luther Burbank published his "New Creations in
Fruits and Flowers" catalog in 1893, revolutionizing the
art of hybridization and furthering a belief in plant will
and sensory perception. Burbank's work is interesting in
connection with Maeterlinck in that it was influenced
by a literary tradition of American writer-naturalists that
includes Henry David Thoreau, John Burroughs, Liberty
Hyde Bailey, and the pioneering botanists, John Bartram
Sr. and William Bartram Jr. The success in Europe of
William Bartram's *Travels* (1791), for instance, influenced
Wordsworth and Coleridge, among others. In 1905 the Vi-
ennese biologist Raoul Francé published *Germs of Mind in
Plants*, while in India at that time Sir Jagadis Chandra Bose
was rigorously shaping his view of the integral relations
between human and plant life.

Among more recent critics, Jacques Vallet (1985) places
Maeterlinck at the head of a line of modern Belgian writers on nature that includes Robert Goffin and Jean de Bos-
chère. Vallet reminds us that Maeterlinck calls for humility
in our perception and understanding of the natural world.
He further notes that in the 1930s Maeterlinck warned far
ahead of modern ecological movements that our survival
depends on our cooperation with nature in all its forms,
and that natural disaster would be the price to pay for our
failure to do so.

What then of Maeterlinck's work in relation to botanical discourse today? We may reasonably argue that
Maeterlinck's essay anticipated a revival of holistic botany
that forms part of a growing ecological consciousness today. The publication in 1973 of *The Secret Life of Plants* by
Peter Tompkins and Christopher Bird renewed the repu-
tations of pioneers like Bose and furthered a vogue for

conversation and intimate behavior between people and plants. Various other writers and naturalists have contributed to this resurgent botanical discourse from such perspectives as literature, horticulture, mathematics, computer graphics, and the visual arts. And though one writer, Michael Pollan, rejects plant intelligence and will in favor of a symbiotic thesis, his meditation on the tulip is clearly in the spirit of Maeterlinck: "We gazed even further into the blossom of a flower and found something more: the crucible of beauty, if not art, and maybe even a glimpse into the meaning of life."[*]

The symbiotic Gaia hypothesis, *pace* Pollan, has certain affinities with Maeterlinck's work albeit minus the Romantic element. Though some of its enthusiasts have attempted to spiritualize or supernaturalize it, this theory of global self-regulation, formulated in the late 1970s by James Lovelock and Lynn Margulis, bases itself on an evolutionary symbiosis among all living organisms, a belief that Maeterlinck might well have found attractive were he still alive today. Indeed, he may have known the coining of *symbiosis* to describe organic coexistence by the German botanist Anton de Bary in 1873. Lovelock's latest vision— of the revenge of Gaia on heedless humankind—reminds us of Maeterlinck's own warnings. Maeterlinck would perhaps also have liked the fact that Gaia (whose name was suggested to Lovelock by William Golding) works against both academic specialization and anthropocentrism in its view of the interdependence of all living species and in its desire to rethink our vital relationship with planet Earth.

---

[*]Michael Pollan, *The Botany of Desire: A Plant's-Eye View of the World* (New York: Random House, 2001) 109.

"What characterizes the most convincing nature writing," suggests Richard Mabey, a leading British naturalist and author of several important botanical studies, "is a willingness to admit both the kindredness and the otherness of the natural world. Its history is thus in part a history of our views about ourselves as a species, part of the quest for the essential characteristics and boundaries of being human."* His description implies that such writing also prompts us to guard our sense of wonder, and to think of our relationship to nature in sympathetic, respectful, and responsible ways. One hundred years after its original appearance, "The Intelligence of Flowers" continues to exemplify this ideal. Even though scientific knowledge, including that of botany, has continued to advance at a great pace since Maeterlinck's time, his essay—erudite, prescient, and eloquent—remains both relevant to our understanding of nature and highly pleasurable to read.

---

*Richard Mabey, *The Oxford Book of Nature Writing* (New York: Oxford University Press, 1995) vii.

The Intelligence of Flowers

The Intelligence of Flowers

I

Here I wish simply to recall several facts known to all botanists. I have made no discovery, and my modest contribution comes down to a few basic observations. It goes without saying that I do not intend to review every proof of intelligence offered to us by plants. Those proofs are ongoing and innumerable, especially among flowers, where the striving of plant-life for light and spirit is at its most focused.

If we allow for some awkward or unfortunate plants and flowers, none is entirely lacking in wisdom and ingenuity. All struggle to accomplish their task; all have the magnificent ambition to overrun and conquer the surface of the earth by thereupon multiplying infinitely the form of existence they represent. To reach this goal, they need, on account of the law that binds them to the soil, to overcome difficulties much greater than those facing the multiplication of animals. For that reason most of them have recourse to ruses, schemes, mechanisms, and traps that in respect, for instance, of mechanics, ballistics, aviation, or observation of insects often predate the inventions and knowledge of mankind.

II

It would be superfluous to redraw the picture of the great systems of floral fertilization: the play of stamens and pistil, the seductiveness of scents, the appeal of harmonious and striking colors, the development of nectar, totally useless to the flower, and which it manufactures only to attract and hold the foreign liberator, the messenger of love, bee, bumblebee, fly, butterfly, moth, which must bring it the kiss of the distant, invisible, motionless lover.

This plant world that strikes us as so tranquil, so resigned, where all seems to be acceptance, silence, obedience, reverence, is on the contrary one wherein the revolt against destiny is at its most vehement and most obstinate. The essential organ, the nourishing organ of the plant, its root, attaches it indissolubly to the soil. If it is difficult to ascertain, among the great laws that overwhelm us, the one that weighs heaviest on our shoulders, for the plant there is no doubt: it is the law that condemns it to immobility from birth to death. So it knows better than we, who fritter our energies, against what it must first arise. And the energy of its obsession, as it rises from the shadows of its roots to organize itself and to blossom in the light of its flower, is an incomparable spectacle. It strains its whole being in one single plan: to escape above ground from the fatality below; to elude and transgress the dark and weighty law, to free itself, to break the narrow sphere, to invent or invoke wings, to escape as far as possible, to conquer the space wherein fate encloses it, to approach another kingdom, to enter a moving, animated world. Is not the fact that it succeeds in doing so as surprising as if we were to succeed in living outside the time assigned us by another destiny or in entering a universe freed from the weightiest laws of matter? We shall see that the flower sets man a prodigious example of insubordination, courage, perseverance, and ingenuity. If we had put into trying to uplift the various inevitabilities that weigh us down—those, for instance, of pain, old age, and death—even half the energy that some tiny flower in our garden has spent, we could be forgiven for thinking our fate would be very different from what it is.

III

In most plants this need for movement, this appetite for space, manifests itself concurrently in both flower and fruit. It is easily explicable in the case of the fruit; or, at any rate, it only reveals a less complex experience and foresight. Contrary to what occurs in the animal kingdom, and because of the terrible law of absolute immobility, the main and worst enemy of the seed is the paternal strain. We are in a strange world where the parents, incapable of moving, know they are condemned to stifle or starve their offspring. All seed that falls at the foot of a tree or plant is lost or will sprout in a hard place. Hence the immense effort to cast off the yoke and conquer space. Hence the marvelous systems of scattering, of propulsion, of aviation, that we find in all parts of the forest and on the plain; among others, merely to mention in passing a few of the most curious: the winged screw or samara of the maple, the bract of the lime, the gliding-machine of the thistle, dandelion, and salsify; the detonating springs of the spurge, the extraordinary squirting pear of the balsam apple, the woolen hooks of the eriophorous plants; and a thousand other unexpected and astonishing mechanisms, for there is, so to speak, no seed that has not invented some method wholly suited to itself for escaping the maternal shadow.

We simply would not believe, if we had not practiced botany, just how much imagination and genius expends itself in all the greenery that delights the eye. Look, for instance, at the pretty seed pod of the scarlet pimpernel, the five valves of the garden balsam, the five spring-loaded capsules of the geranium, etc. While you are at it, do not forget to examine the common poppyhead that one finds at any herbalist's. There is, in that nice big head, a prudence and foresight worthy of the highest praise. We know that it contains thousands of extremely fine, tiny black seeds. The goal is to scatter these seeds in the most adroit manner and as far as possible. If the capsule containing them were to split, fall, or open up underneath, the precious black powder would simply form a useless heap at the foot of the stem. But it can only emerge from apertures pierced high up on the husk. Once ripened, this leans over on its peduncle, sways at the slightest breath of air and, literally with the very same gesture as a sower, scatters the seeds in space.

Shall I speak of the seeds that plan their own dispersal by birds, and which, to entice them, huddle, like the mistletoe, the juniper, the serviceberry, etc., within a sugary husk? There is such reasoning in this, such an understanding of final causes that we hardly dare insist on it for fear of renewing the naive errors of Bernardin de Saint-Pierre. Yet the facts cannot otherwise be explained. The sugary husk is as useless to the seed as is the nectar, which attracts the bees, to the flower. The bird eats the fruit because it is sweet and at the same time swallows the seed, *which is indigestible*. The bird takes off and shortly thereafter rends the seed just as it received it but stripped of its husk and ready to sprout far from the dangers of its place of birth.

IV

But let us get back to simpler schemes. Pick from the roadside, in the first fistful to hand, a blade of some grass or other, and you will catch a small, independent, unflagging, unexpected intelligence at work. Here are two poor creeping plants that you could have encountered a thousand times on your walks, for they may be found everywhere, even in the most barren corners where a pinch of soil has strayed. They are two varieties of wild medic (*Medicago*), two weeds in the humblest sense of that word. One bears a reddish flower, the other a yellow powder puff the size of a pea. To see them slide and hide themselves in the lawn, among the proud grasses, one would never imagine that, well before the illustrious geometrician and physician of Syracuse, they have discovered the amazing properties of the Archimedean screw and attempted to apply it not to the raising of liquids but to the art of flying. So they lodge their seeds in easy spirals, made up of three or four admirably constructed revolutions, reckoning thereby to delay their fall and consequently, with the aid of the wind, to prolong their aerial voyage. One of them, the yellow one, has even perfected the device of the red one by decorating the edges of the spiral with a double row of spikes, with the clear intention of hooking it in passing to either human clothing or animal fleece. Clearly it hopes to ally the advantages of eriophily—that is to say, the scattering of seeds by sheep, goats, rabbits, etc.—to those of anemophily, that is to say, scattering by wind.

What is most touching in this entire huge effort is its uselessness. The poor red and yellow medics have made an error of judgment. Their extraordinary screws are of no use to them. They can only work if they fall from a certain height, from the top of a tall tree or from a grassy knoll; but, constructed at grass level, they have only a quarter-turn to make before hitting the ground. We have here a curious example of the mistakes, the trial and error, the experiments and the minor miscalculations, frequent enough, of nature: for only those who have barely studied it would claim that nature never errs.

In passing, let us note that other varieties of medic (not to mention the clover, another leguminous and papilionaceous plant that is virtually indistinguishable from the one that concerns us here) have not adopted these aeronautical devices and retain the primitive method of the pod. In the case of one of them, the *Medicago aurantiaca*, we may easily grasp the transition from highly elaborate pod to screw. Another variety, the *Medicago scutellata*, rounds off the screw in the form of a bowl. It seems therefore that we are witnessing the stirring spectacle of a kind of invention in progress, the efforts of a family that has not yet worked out its destiny and is seeking the best way of ensuring its future. Is it not perhaps in the course of this research that, having been disappointed by the spiral, the yellow medic adds the spikes or the woolen hooks, telling itself not unreasonably that since its foliage attracts sheep, it is inevitable and proper that sheep should assume responsibility for its procreation? And, in the end, is it not on account of this fresh effort and this bright idea that the yellow-flowered medic is infinitely more widespread than its stronger red-flowered cousin?

V

It is not only in the seed or flower but in the whole plant, stems, leaves, roots that we discover, if we but lower our heads for a moment over their humble work, many traces of a lively and shrewd intelligence. Remind yourselves of the magnificent efforts of thwarted branches seeking the light, or the ingenious and courageous struggle of trees in danger. For my part, I shall never forget the admirable example of heroism set to me by an enormous hundred-year-old laurel the other day in Provence, in the wild and delightful gorges of the Loup, fragrant all over with violets. You could easily read on its tortuous and, so to speak, convulsive trunk all the drama of its tough and difficult life. A bird or the wind, masters of destiny both, had carried the seed to the side of a rock that drops like an iron curtain, and the tree was born there, two hundred meters above the torrent, inaccessible and solitary, among scorching and barren stones. From its very first hours, it had sent its blind roots on a long and painful search for precious water and soil. But this was only the congenital concern of a species that knows the arid Midi. The young stem had to resolve a much graver and more unexpected problem: it set out on a vertical plane, so that its head, instead of reaching for the sky, leaned over the gulf. It had therefore, despite the crushing weight of the branches, to set the first surge straight, stubbornly to bend the frustrated trunk just above the surface of the rock, and thus—like a swimmer throwing back his head—by an incessant will, tension, and contraction, to sustain the heavy crown of leaves rising up into the sky.

From that point on, around that vital knot, were concentrated all the preoccupations, all the energy, all the

8

conscious, free genius of the plant. The monstrous bend, grown abnormally large, revealed one by one the successive anxieties of a kind of thought that knew how to avail itself of the warnings given to it by rain and gale. Year after year the dome of foliage grew weightier, with no other concern but to expand in the heat and light, while a dark canker ate away deeply at the tragic arm maintaining it in space. Then, obeying goodness knows what order of the instinct, two solid roots, two hairy cables, emerging from the trunk at more than two feet above the bend, came to moor it to the granite face. Had they truly been brought forth by distress, or else had they been waiting, perhaps with foresight, since the first days, for the critical hour of danger in order to enhance the value of their assistance? Or was it just a happy coincidence? What human eye will ever capture these silent dramas, too long-lasting for our brief lives?*

VI

Among the plants that offer the most striking proof of initiative, those we might call animated or capable of feelings would merit detailed study. I shall content myself by

---

*Let us put this together with the intelligent action of another root of which Brandis (*On Life and Polarity*) recounts to us the exploits. In getting stuck in the earth, it had met the old sole of a boot; to pass this obstacle, which it was apparently the first of its species to find in its way, it subdivided itself in as many parts as there were holes left by the stitch points, then, once past the obstacle, it reunited itself and knitted together again all its various rootlets, in such a way as to reconstitute a single, homogeneous taproot.

9

recalling the delightful terrors of that sensitive plant we all know, the shrinking mimosa. Other herbs with spontaneous movements are less well known; the *Hedysareae*, in particular, among which the *Hedysarum gyrans* or swaying sainfoin, which bestirs itself in a surprising way. This small leguminous plant, originally from Bengal but often grown in our hothouses, performs a kind of nonstop, intricate dance in honor of the light. Its leaves divide themselves into three leaflets, one broad and terminal, the other two narrow and planted at the base of the first. Each of these leaflets has its own different movement. They live in incessant, rhythmical, and almost chronometrical agitation. They are so sensitive to light that their dance slows down or speeds up according to whether the clouds hide or reveal the chink of sky they face. They are, as we can see, true photometers, and this well before Crookes's invention of the natural otheoscopes.

VII

But these plants, to which we should add the Droseras, the Dionaeas, and many others, are nervous beings already going slightly beyond the mysterious and probably imaginary ridge that separates the plant from the animal kingdom. We have no need to climb that far, and we find as much intelligence and almost as much visible spontaneity at the other end of the world that concerns us, in the lower depths where the plant is barely distinguishable from silt or stone: I refer to the fabulous clan of Cryptogams, observable in detail only beneath a microscope. It is why we pass by this clan in silence, even though the spore play of the mushroom, fern, and above all rough horsetail or scouring rush, is of incomparable delicacy and ingenuity. But among the aquatic plants, living in primeval mud or slime, less secretive wonders are performed. Since the fertilization of their flowers cannot take place under water, each of them has come up with a different system for the pollen to be scattered in the dry zone. Thus the Zosteras, that is to say, the vulgar marine eelgrass with which we stuff mattresses, carefully enclose their flowers in a true divingbell; the waterlilies send their flowers to blossom on the surface of the pond, supporting and nourishing them on an endless peduncle which extends along with any rise in water level. The false waterlily (*Villarsia nymphoides*), with no extendable peduncle, simply lets its own flowers go: they climb to the surface and burst like bubbles. The waterchestnut (*Trapa natans*) equips its own with a kind of air-inflated vessel; they arise and open, then fertilization completed, the air of the vessel is replaced by a mucilaginous liquid heavier than water, and the whole device sinks back into the silt where the fruits will ripen.

The system of the bladderwort is even more intricate. Here is how M. Henri Bocquillon describes it in *The Life of Plants*:

These plants, widespread in ponds, ditches, pools, and puddles of water in peat bogs, are invisible in winter; they rest in the mud. Their extended stem, spindly, trailing, is furnished with leaves reduced to ramified filaments. At the axil of the leaves thus transformed, we notice a kind of small pear-shaped pocket, with an aperture at its pointed upper end. This aperture carries a valve that can only open from the outside in; its sides are finished with ramified hairs; the inside of the pocket is lined with other small secretory hairs that give it an appearance of velvet. When the moment of flowering comes, the little axillary skins fill with air; the more this air tends to escape, the tighter it closes the valve. Eventually it gives the plant great specific buoyancy and guides it to the water's surface. It is only then that those charming small yellow flowers open up, resembling strange little faces with more or less bulbous lips and a palate striped with orange or rust-colored lines. During the months of June, July, and August, they display their fresh colors amid the detritus of vegetation, rising gracefully above the murky water. But fertilization has taken place, the fruit develops, and the roles switch; the surrounding water presses on the valve of the utricles, forces its way in, rushes into the cavity, makes the plant heavier, and compels it to drop again into the mud.

Is it not strange to see collected in this age-old device a few of the most productive and recent human inventions: the play of valves, the pressure of liquids and air, the Archimedean principle studied and utilized? As the author I have just quoted goes on to observe: "The engineer who first attached a flotation device to a sunken vessel could barely have imagined that an analogous process had been in use for thousands of years." In a world that we hold to be unconscious and devoid of intelligence, we imagine at first that the very least of our ideas creates new schemes and relationships. On looking closer at things, it seems extremely likely that we are unable to create anything at all. The last to arrive on this earth, we simply discover what has always existed, and like awestruck children we follow the path that life had made before us. Moreover, it is perfectly natural and reassuring that this be so. But we shall return to this point.

VIII

We cannot leave the aquatic plants without briefly recalling the life of the most fabulous of them all: the legendary eelgrass, a Hydrocharid whose nuptials form the most tragic episode in the love life of flowers.

The eelgrass is a rather unremarkable specimen, with none of the strange grace of the waterlily or of certain sub-aquatic tufts. But we can say that nature has taken pleasure in instilling in it a fine idea. The whole existence of this little plant is spent at the bottom of the water, in a kind of half-sleep, until the hour of the nuptials when it aspires to a new way of life. Then the female flower slowly unfurls the long spiral of its peduncle, rises, emerges, floats, and blossoms on the surface of the pond. From a nearby strain, the male flowers, catching sight of it across the sunlit water, arise in turn, full of anticipation, toward the one that sways, awaits them, calls them to a magical world. But halfway there, they suddenly feel held back: their stem, the very source of their life, is too short; they will never make it into the light, into the one place where union of pistil and stamens can occur.

Is there a crueler oversight or test in all of nature? Imagine the drama of this desire, the homing in on the untouchable, the transparent fatality, and the impossible without visible obstacle!

It would be insoluble, like our own drama upon this earth, but here is where an unexpected element comes into play. Did the males foresee their disappointment? For they always enclose an air bubble in their hearts, like we enclose in our soul a desperate thought of release. It is as if they hesitate for a moment, then with a magnificent effort—the most supernatural that I know of in the annals of insects and flowers—to soar toward happiness, they deliberately break the bond that secures them to life. They tear themselves away from their peduncle and with an incomparable surge, amid pearls of lightheartedness, their petals break the surface of the water. Mortally wounded, but radiant and free, they float momentarily alongside their indifferent fiancées; the union takes place, after which the martyrs drift off to perish downstream, while the already pregnant spouse seals her corolla, where their last gasp survives, rolls up her spiral, and returns to the depths, there to ripen the fruit of the heroic kiss.

Must we spoil this pretty picture, rigorously precise but seen from the bright side, by looking at it from the dark side too? Why not? Sometimes the dark side yields truths as interesting as those from the bright side. This delightful tragedy is only perfect when we consider the intelligence, the aspiration of the species. But if we observe individuals, we often see them acting out this ideal plan awkwardly and incorrectly. Sometimes the male flowers rise to the surface when there are not yet any pistillated flowers in the vicinity. And at other times, when low water permits them easily to reach their companions, they still break their stems no less automatically and uselessly. I maintain here, once again, that the whole genius rests in the species, in life or nature, and that the individual on the whole is stupid. Only in mankind do we find true emulation of the two intelligences, an increasingly precise and active tendency toward a kind of balance that is the great secret of our future.

IX

The parasitic plants also offer us unusual and mischievous sights, such as that astonishing Cuscuta commonly known as the dodder. It has no leaves and its stem needs only reach several centimeters in length for it to abandon its roots deliberately, so as to entwine itself around its chosen victim and into which it sinks its suckers. From then on it lives wholly at the expense of its prey. It is impossible to outwit its insight; it will refuse any sustenance that does not suit it, and it will go in search, as far away as necessary, of any stem of hemp, hop, lucerne, or flax that suits its temperament and taste.

This dodder draws our attention naturally to the creepers, which have highly noteworthy habits of which something should be said. Moreover, those of us who have lived for a while in the countryside have had many opportunities to admire the instinct and kind of vision that directs the tendrils of the Virginia creeper or the morning glory toward the shaft of a rake or spade leaning up against a wall. Move the rake, and the next day the tendril will have fully returned and found it again. Schopenhauer, in his treatise *On the Will in Nature*, in the chapter devoted to the physiology of plants, summarizes this and several other points with a host of observations and experiments that would take too long to list here. I thus refer the reader to Schopenhauer; there he will find information on numerous sources and references. Is there any need for me to add that in over fifty or sixty years these sources have strangely multiplied and that the subject furthermore is virtually inexhaustible?

Among so many different inventions, subterfuges, and precautions, let me also mention, by way of example, the sound judgment of the radiate hyposerid (*Hyposeris radiata*), a small plant with yellow flowers, rather similar to the dandelion, which one finds often on old walls along the Riviera. To ensure the simultaneous spread and stability of its race, it carries at the same time two types of seed. One detaches easily and is equipped with wings to let the wind pick it up, while the other, which does not have wings, remains imprisoned in the inflorescence and is released only when the latter decomposes.

The case of the spiny cocklebur (*Xanthium spinosum*) shows us to what extent certain systems of scattering are well conceived and successfully operated. This cocklebur is a frightful weed bristling with barbaric prickles. Not so long ago it was unknown in Western Europe and nobody, naturally, had thought of introducing it. It owes its conquests to the hooks that complete the capsules of its fruits and catch on animal fleece. Originally from Russia, it came to us in bales of wool imported from the furthest steppes of Muscovy, and we may trace on a map the stages of this great migrant that conquered a new world.

The Italian catchfly (*Silene italica*), a simple little white flower that we find in abundance beneath olive trees, has turned its mind in another direction. Apparently very fearful, very sensitive, it avoids visits from bothersome and cumbersome insects by fitting out its stem with glandular bristles whence oozes a viscous liquid in which so many parasites get trapped that the peasants of the Midi use the plant as a flycatcher in their homes. Moreover, certain kinds of catchfly have ingeniously simplified the system. Since it is especially ants they dread, they have found that it suffices, in order to prevent them passing by, to place a broad sticky ring beneath the node of each stem. It is exactly what gardeners do when they mark a ring of tar around the trunk of apple trees to prevent the ascent of caterpillars.

This will lead us to study the means of defense in plants. M. Henri Coupin, in an excellent popular text, *The Original Plants*, to which I refer the reader desirous of further details, examines some of these strange armaments.

First of all there is the intriguing matter of thorns, upon which subject a student at the Sorbonne, M. Lothelier, has carried out some very interesting experiments, which prove that shade and humidity tend to suppress the prickly parts of plants. On the other hand, the more arid and sun-baked the spot where it grows, the more the plant bristles and multiplies its spikes, as if it understood that being almost the sole survivor of the bare rocks or the scorching sand it must redouble its efforts to defend itself against an enemy that no longer has any choice of prey. Furthermore, it is noticeable that, when grown by man, the majority of thorny plants gradually discard their weapons, giving over the task of their salvation to the supernatural guardian who adopts them in his garden.[*]

Certain plants, among others the borages, replace the thorns with very stiff bristles. Others, like the nettle, add poison to them. Yet others, the geranium, the mint, the rue, etc., impregnate themselves with strong odors to fend

---

[*]Among the plants that have given up defending themselves, the most striking case is that of the lettuce. "In the wild," as the aforementioned author puts it, "if one breaks a stem or a leaf, one sees a white sap escape it, latex, a substance formed from various elements that vigorously defend the plant from attack by slugs. On the contrary, in the cultivated species deriving from the former, there is barely any latex; thus the plant, to the great despair of gardeners, is no longer capable of fighting and lets itself get eaten by slugs." It is worth adding, however, that this latex is hardly ever lacking in other than young plants, instead of which it becomes abundant again when the lettuce begins to "form a head" and when it runs to seed. So it is at the beginning of its life, at the budding of its first and tender leaves, that it will experience the greatest difficulty in defending itself. We might say that the cultivated lettuce somewhat loses its head, if you will excuse the expression, and no longer quite knows where it stands.

off animals. But the strangest of all are those that defend themselves mechanically. I shall but mention the horsetail, which surrounds itself with a veritable armor of microscopic grains of silica. Moreover, nearly all the grasses add lime to their tissues to discourage the voraciousness of slugs and snails.

X

Before touching on the study of the complex devices that cross-fertilization requires, among the thousands of nuptial ceremonies going on in our gardens, let us mention the ingenious ideas of some very simple flowers wherein spouses are born, fall in love, and die in one and the same corolla. We are familiar enough with the type of system: the stamens* or male organs, generally frail and numerous, set themselves up around the robust and patient pistil. "*Mariti et uxores uno eodemque thalamo gaudent*" [Husbands and wives delight in one and the same bed], as Linnaeus so beautifully puts it. But the disposition, form, and habits

---

*At the beginning of this study, which could become the golden book of floral nuptials (the responsibility for which I leave to a greater expert than myself), there is perhaps a point in calling the reader's attention to the faulty, disconcerting terminology used in botany to specify the reproductive organs of plants. In the female organ, the pistil, which includes the ovary, the style, and the stigma which crowns it, all is of masculine gender and all seems virile. On the other hand the male organs, the stamens that top the anthers, have a young girl's name. It is just as well to get this antimony clear in the mind once and for all.

of these organs vary from flower to flower, as if nature had an idea that could not yet be determined or an imagination that makes a point of honor of never repeating itself. Often the ripened pollen falls quite naturally from the top of the stamens onto the pistil, but quite often too, pistil and stamens are of the same size, or the latter are too long, or the pistil is twice their size. Endless attempts to meet up then occur. Sometimes, as in the nettle, the stamens, at the bottom of the corolla, stand cowering on their stem. At the moment of fertilization, the stem slackens like a spring, and the anther or pollen sac that tops it shoots a cloud of dust over the stigma. Sometimes, as in the barberry, so that the nuptials only occur during the best hours of a fine day, the stamens, distanced from the pistil, are held against the walls of the flower by the weight of two moist glands; the sun appears, evaporates the liquid, and the unburdened stamens fall on the stigma. Elsewhere it differs: thus in the primrose the females are by turns longer or smaller than the males. In the lily, tulip, etc., the spouse, too slender, does what she can to gather and concentrate the pollen. But the most original and fantastic system is that of the rue (*Ruta graveolens*), a quite foul-smelling medicinal herb, of the disreputable group of emmenagogues. The stamens, still and docile in the yellow corolla, wait, lined up in a circle around the large squat pistil. At the nuptial hour, obeying an order from the female, who seems to make a kind of nominal appeal, one of the males approaches and touches the stigma, followed by the third, the fifth, the seventh, and the ninth male until the whole odd row is done. Then, in the even row, it is the turn of the second, the fourth, the sixth, etc. It is truly love on demand. This flower that can count seemed so extraordinary to me that I did not at first believe the botanists and tried more than once to verify its sense of number before confirming it. I have noticed that it rarely makes a mistake.

It would be overdoing it to multiply these examples. A simple stroll through field or wood will permit a thousand observations of this point quite as curious as those that the botanists report. But before bringing this section to a close, I must insist on pointing out one more flower; not that it bears witness to a truly extraordinary imagination, but for the delightful and easily perceptible grace of its gesture of love. It is the fennel flower (*Nigella damascena*) whose common names have a certain charm: love-in-a-mist, devil-in-the-bush, beauty-with-flowing-locks, etc.—such bold and touching efforts of popular poetry to describe a little plant it likes. We find this plant growing wild in the Midi, at the roadside and beneath the olives, and further north it is often found growing in rather old-fashioned gardens. The flower is of a delicate blue, simple like a primal floweret, and the "flowing locks" are the intertwined leaves, slender and light, that surround the corolla with a "bush" of diaphanous foliage. At the source of the flower, the five extremely long pistils stand tightly grouped in the center of the sky-blue crown, like five queens clothed in green gowns, haughty and inaccessible. Around them in a countless crowd gather their hopeless lovers, the stamens, which fail even to come up to their knees. Then at the heart of this palace of turquoises and sapphires, in the bliss of a summer's day, begins the silent unending drama, that of an impotent, useless, static waiting. But hours go by, years in the life of the flower; its glory fades, its petals fall, and the arrogance of the great queens appears finally to be sinking beneath the weight of life. At a given moment, as if they were obeying a secret and irresistible word of love that deems the test sufficient, all of them together lean backward in a concerted and symmetrical movement, comparable to the harmonious parabolas of a five-pronged fountain falling back into its basin, and come gracefully to take, from the lips of their humble suitors, the golden powder of the nuptial kiss.

XI

As we see, the unexpected abounds here. A great book, therefore, is waiting to be written on the intelligence of plants, as Romanes did on the intelligence of animals. But this brief sketch is by no means intended to become a manual of that kind; I wish merely to draw attention to several interesting events taking place around us in this world wherein we believe ourselves rather too vainly to be the privileged ones. These events have not been carefully selected but are offered as simple examples from random observations and circumstances. Moreover, in these brief notes I try to concern myself above all with the flower, for in it shines forth the greatest wonders. For the time being I am omitting carnivorous flowers—sundews, various kinds of pitcher plants, etc., that border on the animal kingdom and would require a special, elaborate study—to concentrate on the flower in the true sense of the word, on that which we consider insentient and inanimate.

So as to separate facts from theories, let us speak of the flower as if it had foreseen and conceived of its achievement in a human way. We shall see later what it must still be credited with and what ought to be taken away from it. For the moment there it is alone onstage, like a magnificent princess endowed with reason and will. We cannot deny that it seems to have been thus imbued; to strip it of those attributes we must turn to extremely vague hypotheses. So there it is, immobile on its stem, sheltering the reproductive organs of the plant in a dazzling tabernacle. It seems as if it has only to let itself accomplish the mysterious union of stamens and pistil in the heart of this tabernacle of love. And many flowers allow this to happen. But for many others a terrible threat looms large, posing the generally insoluble problem of cross-fertilization. Following what countless and forgotten experiments have they recognized that self-fertilization, that is to say fertilization of the stigma by pollen fallen from the anthers surrounding it in the same corolla, brings with it the rapid degeneration of the species? They have recognized nothing, nor profited from any experiment, we are told. The force of things simply and gradually has eliminated seeds and plants weakened by self-fertilization. Soon only those survived in which some kind of anomaly—for instance, the exaggerated length of the pistil being inaccessible to the anthers—prevented them from fertilizing themselves. These exceptions being the only ones to survive a thousand strange incidents, heredity finally arranged for chance to do its work, and the normal type disappeared.

XII

Further on we shall see what these explanations reveal. For the moment let us go out again into the garden or the open countryside, in order to study more closely two or three unusual inventions of the flower's genius. And already, without straying far from the house, here, haunted by bees, is a fragrant cluster inhabited by a highly skilled mechanic. No one, not even the least countrified of us, can fail to be acquainted with the wholesome sage. An unpretentious labiate, it bears a very modest flower that opens vigorously, like a starving gob, to grab the sunrays as they pass by. Moreover, a great number of varieties exist, which—curious detail, this—have not all adopted or brought to the same perfection the system of fertilization we are about to examine.

But I shall only concern myself here with the most common form of sage, that one at present, as if to celebrate the rite of spring, covering every wall of my olive-tree terraces with violet drapery. I can assure you that the balconies of great marble palaces, where kings are expected to appear, have never been more happily, luxuriously, or fragrantly adorned. We almost imagine catching the scents from the light of the sun at its height, when midday strikes.

To go into detail, the stigma or female organ of the flower is enclosed in the upper lip, which forms a kind of hood, where the two stamens or male organs are also to be found. In order to prevent them fertilizing the stigma that shares the same nuptial pavilion, this stigma is twice their length, so that they have no hope of reaching it. Furthermore, to avoid any accidents, the flower has made itself protandrous, that is to say, the stamens ripen before the pistil, so when the female is ready to conceive, the males have already disappeared. An external force must therefore intervene to accomplish the union by transporting a foreign pollen to the abandoned stigma. A certain number of flowers, the anemophilous ones, leave this task to the wind. But the sage—and this is the more general case—is entomophilous, that is to say, it loves insects and relies solely on their collaboration. Moreover, it is fully aware—for it knows plenty of things—of living in a world where it pays not to expect any sympathy, any charitable act. So it does not waste its time in making useless appeals to the indulgence of the bee. The bee, like everything that struggles against death on this earth, lives only for itself and its own kind, and is quite unconcerned to render service to the flowers that sustain it. How then to force it, despite itself, or at least without its knowledge, to carry out its matrimonial duty? Here then is the superb love trap conjured up by the sage: right at the back of its tent of violet silk, it distils a few drops of nectar; that is the bait. But blocking the flow of the sugary liquid are two parallel stems that stand upright rather like the shafts of a Dutch drawbridge. At the very tip of each stem is a large blister, the anther, teeming with pollen; below, two smaller blisters serve as counterweights. When the bee enters the flower, in order to get at the nectar, it has to push the small blisters with its head. The two stems, which pivot on an axis, suddenly topple, and the upper anthers come into contact with the sides of the insect, covering them in pollen dust.

As soon as the bee has left, the spring pivots return the mechanism to its original position, and everything is ready to go again on the next visit.

However, that is only part one of the drama: the sequel unfolds in another setting. In a neighboring flower, where the stamens have just wilted, the pistil awaiting pollen enters the stage. It slowly issues from the hood, elongates itself, leans over, curves downward, and forks, in its turn blocking entry to the pavilion. Going for the nectar, the head of the bee passes freely beneath the hanging fork, but this fork scuffs the bee's back and sides, exactly at the points where the stamens touch. The double-cleft stigma greedily absorbs the silvery dust and the impregnation goes ahead. Furthermore, by introducing into the flower a piece of straw or the tip of a matchstick, it is easy to get the device to stir and to realize the striking and marvelous combination and precision of all its movements.

The varieties of sage are numerous, about five hundred in all, and I shall spare you most of their scientific names which are seldom elegant: *Salvia pratensis, officinalis* (the one in our vegetable plots), *horminum, horminoides, glu-tinosa, sclarea, roemeri, azurea, pitcheri, splendens* (the magnifi-cent scarlet sage of our hanging baskets), etc. Perhaps there is not a single one that has not modified in some detail the device we have just examined. Some—and this is, I believe, a dubious improvement—have doubled, occasion-ally tripled the length of the pistil, in such a way that it not only emerges from the hood, but curves upward in a full plume in front of the entrance to the flower. Thus they avoid the faint danger of fertilization of the stigma by the anthers located within the same hood, but on the contrary, it may happen that if the protandry be inexact, the bee, on leaving the flower, deposits on the stigma the pollen of the very anthers with which the stigma coexists. Others, in the leveraging movement, make the anthers diverge further, and in this way they strike the sides of the creature with greater certainty. Others, finally, have failed yet to arrange and adjust all the parts of the device. I find, for instance, not far from my violet sage, close by the well, beneath a cluster of oleanders, a family of white flowers tinted with pale lilac. There we find neither plan for nor sign of a lever. The stamens and the stigma haphazardly clutter the cen-ter of the corolla. Everything there seems disorganized and given over to chance. I do not doubt it is possible, for those who collect the many varieties of this labiate, to piece together the whole history, to follow all the stages of invention, from the primal disorder of the white sage I am looking at now to the latest perfections of the medici-nal sage. What does this all mean? Is the system still being developed by the aromatic tribe? Are we still in the phase of fine tuning and testing, as with the Archimedean screw in the sainfoin family? Have we still to acknowledge unanimously the excellence of the automatic lever? All therefore not being immutable or preordained, would one be arguing and experimenting in a world we deemed fatally, organically habitual?*

XIII

Be that as it may, the flower of most varieties of sage offers an attractive solution to the great problem of cross-fertilization. But just as, among men, a new invention is immediately taken up, simplified, improved by a crowd of indefatigable minor researchers, so in what we might call the "mechanical" world of flowers, the patent of the sage has been worked out and, in many respects, strangely perfected.

---

*For several years now I have been carrying out a series of experiments with the hybridization of the sage, artificially fertilizing, after the customary precautions to exclude any intervention by the wind or by insects, a variety whose floral mechanism has reached a high state of perfection, with the pollen of a highly retrograde variety, and vice versa. My observations are not yet sufficiently numerous for me to go into detail here. Nonetheless, it seems that a general law already begins to emerge from them, namely, that the retrograde sage willingly adopts the improvements of the advanced variety, rather than the latter assuming the defects of the former. In this we get a curious enough glimpse into nature's procedures, habits, preferences, and taste for the very best. But these experiments are inevitably slow and lengthy because of the time taken in gathering the different varieties, in making necessary checks and double checks, etc. It would therefore be premature to draw the slightest conclusion from them.

A fairly ordinary member of the wort family, the small louse-wort (*Pedicularis sylvatica*), which you have surely come across in the shady parts of copses and heaths, has brought some extremely ingenious modifications to the process. The form of the corolla is just about the same as that of the sage; the stigma and the two anthers are located all three in the upper hood. Only the small moist tip of the stigma overshoots the hood, while the anthers remain total captives within. In this silken tabernacle, the organs of the two sexes are cramped together and even in direct contact; nonetheless, thanks to a system quite different to that of the sage, self-fertilization is totally impossible. In fact, the anthers form two blisters filled with powder; these blisters, each of which has only one opening, are juxtaposed in such a way that if these openings coincide, they mutually close each other off. They are forcibly kept within the hood, on their folded, springy stems, by two kinds of teeth. The bee or bumblebee that goes into the flower to draw the nectar from it must open up these teeth; as soon as they are freed, the blisters shoot up, fall outside, and drop upon the insect's back.

But the genius and foresight of the flower does not stop there. As H. Müller points out, being the first to study in full the prodigious mechanism of the lousewort:

> If the stamens were to strike the insect while retaining their relative positions, not one single grain of pollen would emerge, since their orifices close each other off. But a device as simple as it is ingenious resolves the problem. The lower lip of the corolla, instead of being symmetrical and horizontal, is irregular and slanting,

so that one side is several millimeters higher than the other. The bumblebee positioned on it must also be leaning. Its head consequently strikes the projections of the corolla one after another. Thus the release of the stamens occurs in succession as well, so that first one, then the other, their orifices freed, strikes the insect and sprinkles it with the fertilizing dust.

When the bumblebee then passes to another flower, it inevitably fertilizes it, for—this is a detail purposely omitted—what it encounters at the very first moment of pushing its head into the entrance of the corolla is the stigma, which brushes it right at the spot where it goes a moment later to be marked by the impact of the stamens, at that exact spot where it has already been touched by the stamens of the flower it has just left.

XIV

We might multiply these examples indefinitely, each flower having its idea, its system, its acquired experience that it turns to advantage. In examining closely their little inventions, their various procedures, we recall those tremendous exhibitions of machine tools, where the mechanical genius of man reveals its range of resources. But our mechanical genius dates from yesterday, whereas floral mechanics have been working for thousands of years. When flowers first appeared on this earth, they had no models around them to imitate; they had to come up with everything from within themselves. From a time when we were still waving clubs, aiming bows, swinging flails, to fairly recent times when we thought up the spinning wheel, the pulley, the hoist, the ramrod; at the time—it was last year, so to speak—when our masterpieces were the catapult, the clock, and the loom, the sage had fashioned the pivoting shafts and counterweights of its precise swing, and the lousewort its blisters sealed off as though for a scientific experiment, its repeatedly triggering springs, and its scheme of inclined planes. Who then, less than one hundred years ago, had any idea of the properties of the screw that the maple and the lime have used since the birth of trees? When will we reach the stage of fashioning a parachute or a flying machine as firm, as light, as subtle and as safe as that of the dandelion? When will we discover the secret of carving in a material as fragile as the silk of petals a spring as powerful as that which hurls into space the golden pollen of the Spanish broom? And as for the balsam apple or squirting cucumber, whose name I gave at the beginning of this little study, who will tell us the mystery of its miraculous strength? Do you know the balsam? It is a humble cucurbit, found all along the Mediterranean coast. Its fleshy fruit, which resembles a small cucumber, is endowed with inexplicable vitality and energy. You have but to touch it, at the moment of its maturity, for it to detach itself suddenly from its peduncle in a convulsive contraction and to shoot through the opening created by the wrench, mingled with many seeds, a mucilaginous jet of such exceptional force that it carries the seed four or five meters away from the natal plant. The action is as extraordinary as if we had succeeded, relatively speaking, in emptying ourselves in one spasmodic movement and had shot all our organs, innards, and blood half a kilometer from our skin and skeleton. Furthermore, a large number of seeds make use of ballistic methods and sources of energy that remain more or less unknown to us. Recall, for instance, the cracking report of the rape and the broom. But one of the great masters of vegetal artillery is the spurge. The spurge is an indigenous euphorbiate, a huge and decorative "weed" that often surpasses a man in size. Right now I have on my table a branch of spurge steeping in a glass of water. It bears triple-cleft, greenish berries that enclose the seeds. From time to time one of these berries bursts with a loud report, and the seeds, gifted with a prodigious initial velocity, strike the furniture and walls on every side. If one of them hits you in the face, you would think you had been stung by an insect, such is the extraordinary penetrative force of these tiny seeds of pinhead size. Examine the berry, look for the springs that move it, but you will not find the secret of this force; it is as invisible as that of our nerves. The Spanish broom (*Spartium junceum*) has not only pods but spring-loaded flowers too. Perhaps you have noticed this admirable plant. It is the finest representative of that powerful family of brooms, fiercely clinging to life, poor, sober, robust, dreading no earth, no challenge. Along the pathways and in the mountains of the Midi, it forms huge tufted balls, occasionally three meters high, which, between May and June, are covered with a magnificent bloom of pure gold. Its scents, mingling with those of its regular neighbor, the honeysuckle, spread beneath the fury of a harsh sun such delights that can be defined only by evoking celestial dews, Elysian springs, the freshness and limpidity of stars in the depth of blue grottoes.

The flower of this broom, like that of all the papilionaceous *Leguminosae*, resembles the flower of our garden pea; its lower petals, welded together like a galley's ram, hermetically seal the stamens and the pistil. As long as it remains unripe, the exploring bee finds it impenetrable. But as soon as the moment of puberty arrives for the captive fiancés, the ram bends under the weight of the alighting insect, and the golden chamber explodes voluptuously, propelling far and wide, with great force, over the visitor, over the nearby flowers, a cloud of luminous dust, which a broad petal arranged as a canopy casts down, with an excess of caution, upon the stigma to be impregnated.

XV

I refer those who may care to study all these problems to the works of Christian Konrad Sprengel, who was the first, in his unusual work, *The Secret of Nature Revealed* (1793), to analyze the functions of the different organs in orchids; then to the books of Charles Darwin, H. Müller of Lippstadt, Hildebrand, Delpino the Italian, Sir William Hooker, Robert Brown, and many others.

We shall find the most perfect and most harmonious manifestations of plant intelligence among the orchids. In these strange and convoluted flowers, the genius of the plant attains its highest point and with an unusual brilliance pierces the wall that divides the kingdoms. Moreover, we must not allow this name "orchid" to mislead us to believe that this concerns only rare and precious flowers, those queens of the hothouse that seem to claim the attention of the goldsmith rather than the gardener. Our wild, indigenous flora, including all our modest "weeds," comprise more than twenty-five species of orchid, among which we find the most ingenious and most complicated specimens. It is these which Charles Darwin studied in his book, *On the Fertilization of Orchids by Insects*, which is the amazing history of the most heroic efforts of the soul of the flower. There is absolutely no question of summarizing here, in a few lines, this prolific and magical biography. Nonetheless, since we are concerned with the intelligence of flowers, it is necessary to give a general idea of the mental processes and habits of that which surpasses all others in the art of compelling the bee or the butterfly to do exactly what it wants, in the form and time prescribed.

XVI

Without drawings it is not easy to get you to understand the extraordinarily complex mechanism of the orchid. I shall try anyway to give a general idea of it with the aid of more or less approximate comparisons, while avoiding as far as possible the use of technical terms such as *retinaculum*, *labellum*, *rostellum*, etc., which give rise to no specific image in the minds of those unfamiliar with botany.

Let us take one of the most widespread orchids in our regions, the *Orchis maculata*, for instance, or rather, for it is a little larger and thus easier to observe, the *Orchis latifolia*, the broad-leaved *Orchis*, commonly called Whitsuntide flower. It is a perennial plant reaching thirty to sixty centimeters in height, commonly enough found in woods and water meadows, and bearing a thyrse of little pinkish flowers that bloom in May and June.

The typical flower of our orchids represents fairly precisely a fantastic and open-mouthed Chinese dragon's head. The highly elongated and hanging lower lip, in the form of a toothed or jagged apron, serves as a base or resting place for the insect. The upper lip rounds in a kind of hood that shelters the vital organs; while behind the flower, beside the peduncle, drops a kind of prow or long pointed horn that contains the nectar. In most flowers, the stigma or female organ is a small and fairly sticky tuft that awaits the coming of the pollen patiently, at the end of a fragile stem. In the orchid, this classic installation has grown unrecognizable. Deep inside the mouth, at the spot occupied in the throat by the uvula, we find two tightly welded stigmas, above which rises a third stigma modified into an extraordinary organ. At its top it carries a kind of pouch, or more precisely, a kind of stoup known as the rostellum. This stoup is full of a sticky fluid in which soak two tiny pellets whence emerge two short stems laden at their tops with a meticulously tied up packet of pollen grains.

Let us see now what happens when an insect enters the flower. It alights on the lower lip outspread to receive it and, drawn by the scent of the nectar, tries to get at the horn that contains it deep within. But by design, the way through is very narrow, and as it moves forward, the insect's head cannot avoid striking the stoup. Immediately this stoup, sensitive to the slightest blow, tears open along a convenient line, exposing the two pellets coated with sticky fluid. On making direct contact with the visitor's skull, these pellets attach themselves to it and stick solidly, so that, when the insect leaves the flower, it carries them away and, with them, the two stems they support ending in the tied-up packets of pollen. So there we have the insect capped with two upright horns in the shape of a champagne bottle. Unconscious artisan of difficult work, it next visits a neighboring flower. If its horns remain stiff, they will simply strike with their pollen packets those other packets whose feet are soaking in the watchful stoup, and nothing will issue from the mingling of pollens. Here the genius, experience, and foresight of the orchid stand out. It has calculated down to the last second the time required by the insect to suck the nectar and move to the next flower, and it has figured this out to be on average a thirty-second interval. We have seen that the packets of pollen are borne on two short stems inserted into the sticky pellets; now, at the point of insertion, we find, beneath each stem, a small membranous disk whose sole function is, after thirty seconds, to contract and to fold each of these stems, so that they describe a ninety-degree arc. It is the result of a fresh calculation, on this occasion not in time, but in space. The two horns of pollen that cap the nuptial messenger are now horizontal and pointing in front of its head, so that, when it enters the next flower, they will strike precisely against the two welded stigmas beneath the overhanging stoup.

That is not all, and the genius of the orchid has not yet exhausted its foresight. The stigma that takes the blow from the pollen packet is coated with a sticky substance. If this substance were as thoroughly adhesive as that contained by the stoup, the pollen mass, its stems broken, would get totally stuck in it and would remain fixed there, and its destiny would be complete. That must not be; it is vital that the pollen's chances be not exhausted in a single venture, but rather be multiplied to the maximum extent. The flower that counts the seconds and measures the lines is even a chemist as well and distils two types of gum: one extremely adhesive and which hardens immediately on contact with air, in order to glue the pollen horns to the insect's head, the other greatly diluted, for the work of the stigma. This latter is just tacky enough to unravel or slightly disturb the taut and elastic threads that envelop the pollen grains. Some of these grains stick to it, but the pollen mass is not destroyed, and when the insect goes on to visit other flowers, it will continue its fertilizing work almost indefinitely.

Have I expounded the entire miracle? No, we must still attend to many neglected details, among others, the movement of the little stoup, which after its membrane has split to unveil the sticky pellets, immediately lifts up its lower side in order to keep in good condition in the liquid glue the pollen packet that the insect may not have carried off. There is every reason to note also the very unusually combined divergence of the pollen stems on the insect's head, as well as certain chemical precautions common to all plants, for very recent experiments by Gaston Bonnier seem to prove that every flower, in order to keep its species intact, secretes toxins that destroy or sterilize any foreign pollen. That is just about all we can see, but here, as in everything, the really great miracle begins where our gaze comes to an end.

XVII

I have just now found, in a wild corner of the olive grove, a superb head of lizard orchid (*Loroglossum hircinum*), a variety that, for no apparent reason (perhaps it is extremely rare in England), Darwin has not studied. Of all our indigenous orchids, it is undoubtedly the most remarkable, most fantastic, most astonishing. If it had the size of an American orchid, we could confirm that no more fanciful plant exists. Imagine a thyrse, along the lines of the hyacinth, but a little taller. It is symmetrically adorned with vicious three-cornered flowers of a greenish white stippled with pale violet. The lower petal, decorated at its source with bronzed caruncles, with Merovingian mustaches, and with ominous lilac buboes, extends endlessly, crazily, improbably, in the shape of a twirled ribbon, of the color of a drowned person whose corpse has been in a river for a month. The whole thing, which conjures up an impression of the worst illnesses and seems to blossom in heaven knows which lands of ironic nightmares and evil spells, gives off an awful stink, as of a billy goat, which spreads far and wide revealing the presence of the monster. I am indicating and describing this foul-smelling orchid in this way, because it is common enough in France, is easily recognizable, and lends itself very well, on account of its size and the distinctness of its organs, to the experiments we wish to do with it. In fact, we need only insert the tip of a matchstick into the flower, pushing it carefully to the bottom of the nectary, in order to view, with the naked eye, all the stages of fertilization. Grazed along the way, the pouch or rostellum drops down, revealing a small sticky disk (the lizard orchid has but one) that supports the two pollen stems. The moment this disk violently clutches the tip of the wood, the two boxes that enclose the pollen pellets open lengthwise, and when we retract the matchstick, its tip is solidly capped by two stiff and diverging horns, each ending in a golden ball. Unfortunately, we cannot enjoy here, as in the experiment with the *Orchis latifolia*, the charming spectacle offered by the precise and gradual leaning of the two horns. Why do they not drop down at all? It suffices to push the capped matchstick into a nearby nectary to confirm the superfluousness of this movement, the flower being much larger than that of the *Orchis maculata* or *latifolia* and the nectar horn placed in such a way that, when the insect laden with pollen enters it, this mass arrives exactly at the level of the stigma to be impregnated.

Let us add that it is vital to the success of the experiment to choose a fully ripened flower. We do not know when this is the case, but the insect and the flower know, for the latter does not invite its necessary guests, by offering them a drop of nectar, until its entire apparatus is ready to work.

XVIII

There we have the basis of the fertilization system adopted by our indigenous orchids. But each species, each family, modifies it, perfects its details according to its own experience, psychology, and special preferences. The *Orchis* or *Anacamptis pyramidalis*, for instance, one of the most intelligent, has added to its lower lip or labellum two small ridges that guide the proboscis of the insect toward the nectar and compel it to accomplish everything expected of it. Darwin quite rightly compares this ingenious accessory to the instrument we use sometimes to guide a thread through the eye of a needle. Another interesting improvement: the two little pellets that carry the pollen stems and soak in the stoup are replaced by a single sticky disk, in the form of a saddle. If, following the path to be taken by the insect's proboscis, we insert a needle point or a hog's bristle into the flower, we see very clearly the advantages of this simpler and more practical device. As soon as the bristle has brushed the stoup, the latter breaks along a symmetrical line, revealing the saddle-shaped disk, which immediately attaches itself to the bristle. Pull this bristle out quickly and you will have just enough time to catch the attractive action of the saddle which, seated on the bristle or needle, folds its two lower wings in such a way as to grip tightly the object supporting it. The purpose of this movement is to strengthen the adhesive power of the saddle and especially to ensure more exactly than in the broad-leaved orchid the necessary divergence of the pollen stems. As soon as the saddle has hugged the bristle and as the stems planted in it, drawn apart by its contraction, forcibly diverge, the second movement of the stems begins with a leaning toward the tip of the bristle, in the same way as in the orchid we studied earlier. These two combined movements are completed in thirty to thirty-four seconds.

XIX

Is it not exactly in this way, by trifles, by repetitions, by successive alterations, that human inventions progress? We have all been following, in the latest of our mechanical industries, the minute but nonstop improvements in ignition, in carburetion, in clutch mechanism, in gear transmission. We could truly say that ideas come to flowers in the same way they come to us. Flowers grope in the same darkness, encounter the same obstacles and the same ill will, in the same unknown. They know the same laws, same disappointments, same slow and difficult triumphs. It seems they have our patience, our perseverance, our self-love; the same finely tuned and diversified intelligence, almost the same hopes and the same ideals. Like ourselves, they struggle against a vast indifferent force that ends by helping them. Their inventive imagination follows not only the same cautious and painstaking methods, the same tiresome little pathways, narrow and twisting, but also takes unexpected leaps forward that suddenly finalize an uncertain brain-wave. It is thus that a family of great inventors among the orchids, a rich and strange American family, the *Catasetidae*, abruptly overturned with a daring idea a number of habits that doubtless seemed to it too primitive. First of all, the separation of the sexes is absolute; each has its own special flower. Next, the pollen complex, or mass or packet of pollen, no longer soaks its stem in a stoup full of gum, lying in wait there, a little inertly and in any case stripped of initiative, for the stroke of luck that will attach it to the insect's head. It is folded back on a powerful spring in a kind of cell. Nothing special attracts the insect to this cell. Nor have the superb *Catasetidae* reckoned, like the common orchids, on this or that movement of the visitor; a controlled and precise movement, if you will, but nonetheless one by chance. No, the insect no longer enters a mere admirably engineered flower, but an animated and literally sensitive one. Barely has it placed itself on the magnificent forecourt of bronzed silk than long and nervous feelers, that it cannot avoid brushing, sound the alarm throughout the edifice. The cell—wherein the pollen mass, split into two packets, has been held captive on its folded pedicle supported by a large sticky disk—is immediately torn asunder. Abruptly released, the pedicle straightens like a spring, pulling along with it the two pollen packets and the sticky disk, which are violently ejected. Following a curious ballistic calculation, the disk is always launched first, and strikes the insect, adhering to it. Stunned by the blow, the insect thinks only of escaping the aggressive corolla as quickly as possible and taking refuge in a nearby flower. That is exactly what the American orchid had in mind.

XX

Shall I also point out the curious and practical simplifications that another family of exotic orchids, the *Cypripedeae*, introduces into the general system? Let us remember the circumvolutions of human inventions; we have here an amusing counterproof. A fitter in the workshop, a student assistant in the laboratory, says one day to his superior: "What if we just try to do the opposite? What if we reverse the movement? What if we invert the mixture of fluids?" We try the experiment, and from the unexpected the unknown suddenly issues. It would not be too hard to believe that the *Cypripedeae* have held similar discussions among themselves. We all know the *Cypripedium* or lady's slipper; with its huge jutting chin, its fierce and venomous demeanor, it is the most characteristic of our hothouse flowers, the one that seems to us to be the typical orchid, so to speak. The *Cypripedium* has boldly suppressed the entire delicate and complicated device of the spring-loaded pollen packets, the diverging stems, etc. Its clog-like chin and a sterile, shield-shaped anther block the entrance in such a way as to compel the insect to pass its proboscis over two small heaps of pollen. But this is not the crucial point; the entirely unexpected and abnormal fact is that, contrary to what we have established in all the other species, it is no longer the stigma, the female organ, that is sticky, but the pollen itself. Its grains, instead of being pulverulent, are covered with a coating so glutinous that it may be stretched and drawn out in strings. What are the advantages and drawbacks of this new arrangement? We might fear that the pollen transported by the insect will become attached to any other object but the stigma; on the other hand, the stigma is exempt from secreting the fluid destined to sterilize all foreign pollen. In any case, this problem would require a special study. In the same way, there are patents whose usefulness is not immediately apparent to us.

XXI

To finish with this strange tribe of orchids, it remains only for me to say a few words about a secondary organ that sets the whole mechanism going: I mean the nectary. It has for that matter been the object, on the part of the genius of the species, of research, of endeavors, and of experiments as intelligent and as varied as those that endlessly modify the economy of the vital organs.

The nectary, as we have seen, is in principle a kind of long prow, a long pointed horn that opens up deep in the bottom of the flower, beside the peduncle, and acts more or less as a counterpoise to the corolla. It contains a sugary fluid, the nectar, which feeds butterflies, beetles, and other insects and which is made into honey by the bee.

Its responsibility therefore is to attract the indispensable guests. It has conformed to their size, their habits, their tastes; it is always placed so that they cannot insert or withdraw their proboscis without scrupulously and successively performing all the rituals prescribed by the organic laws of the flower.

We already know enough about the whimsical character and imagination of the orchids to foresee that here, as elsewhere—and even more than elsewhere, for the more flexible organ lends itself much better to this task—their inventive, practical, perceptive, and finicky spirit is given free rein. One of them, for instance, the *Sarcanthus teretifolius*, probably failing to develop a sticky fluid that would harden quickly enough to stick the pollen packet to the insect's head, has solved the problem by taking pains to detain as long as possible the proboscis of the visitor within the narrow passages leading to the nectar. The labyrinth it has plotted is so complex that Bauer, Darwin's skillful illustrator, had to admit defeat and abandon his effort to reproduce it.

There are some which, starting from the excellent principle that all simplification is improvement, have boldly done away with the nectar horn. They have replaced it with various fleshy outgrowths, strange and evidently succulent, which the insects nibble. Need I add that these outgrowths are always placed in such a way that the guest who feasts on them cannot avoid setting the entire pollen mechanism in motion?

XXII

But, without lingering over a thousand small and extremely varied ruses, let us end this fairy tale by studying the lure of the *Coryanthes macrantha*. In fact, we no longer quite know what kind of being we are dealing with. The astonishing orchid has come up with this: its lower lip or labellum forms a kind of large jar into which continually fall drops of almost pure water, secreted by two horns located above; when this jar is half-full, the water flows out on one side via a drainpipe. All this hydraulic installation is already quite remarkable, but here is where the unnerving, I might say almost devilish, aspect of the scheme comes into play. The liquid secreted by the horns and gathered in the satin bowl is not nectar and is in no way meant to attract the insects; it has a much more delicate mission, in the truly Machiavellian scheme of this strange flower. The unsuspecting insects are tempted to enter the trap by sugary scents diffused by the aforementioned fleshy outgrowths. These outgrowths are located above the jar, in a kind of chamber with access via two lateral openings. The big visiting bee—the flower being huge, it invariably seduces only the heaviest *Hymenopterae*, as if the others felt some kind of shame in entering such vast and sumptuous chambers—the big bee starts to nibble the tasty caruncles. If it were alone, it would leave quietly after its meal, without even brushing the jar of water, the stigma, and the pollen; and none of the necessary things would be done. But the wise orchid has been observing life going on around it. It knows that bees form an innumerable, greedy, and busy people, who emerge in their thousands when the sun is out, that all it takes for them to come running en masse to the feast prepared beneath the nuptial tent is a scent to thrill like a kiss on the threshold of an opening flower. So here we have two or three gatherers in the sugary chamber; the place is cramped, the surfaces are slippery, the guests violent. They crowd in, jostling one another, so much so that one of them always ends up falling into the jar awaiting it beneath the treacherous meal. There it finds an unexpected bath, conscientiously wets its beautiful diaphanous wings, and despite tremendous efforts, fails to take off again. The astute flower watches for this. In order to get out of the magical jar, only one opening exists, the drainpipe that serves as an overflow from the reservoir. It is precisely wide enough to allow passage to the insect whose back first of all touches the sticky surface of the stigma, then the viscid glands of the pollen mass that await it along

47

the vault. So it escapes, laden with the adhesive powder, enters a nearby flower, where yet again the drama of the meal, the crush, the fall, the bath, and the escape is reenacted, and inevitably places the imported pollen in contact with the greedy stigma.

Here then is a flower that knows and exploits the passions of insects. We cannot simply claim that all this is no more than many rather fanciful interpretations; no, the facts come from precise and scientific observation, and it is impossible to explain otherwise the use and arrangement of the various organs of the flower. We must accept the evidence. This incredible and efficient ruse is all the more surprising in that here it aims not to satisfy the need to eat, urgent and immediate, that stimulates the dullest witted intelligences; it has in view only a distant ideal: the propagation of the species.

But why, we shall ask, these fantastic complications that serve in the end only to increase the dangers of chance? Let us not rush to judge and reply. We know nothing of the plant's own reasons. Do we know the obstacles it encounters in respect of logic and simplicity? Do we know thoroughly even one of the organic laws of its existence and growth? Someone seeing us from the heights of Mars or Venus, as we strive to conquer the air, would wonder the same: why those misshapen and monstrous machines, those balloons, those airplanes, those parachutes, when it would be so simple to imitate the birds by fitting out the arms with a pair of perfectly satisfactory wings?

XXIII

To these proofs of intelligence, man's rather puerile vanity offers the traditional objection: yes, they create marvels, but those marvels remain the same for ever. Each species, each variety has its system and, throughout the generations, brings no noticeable improvement to it. It is very true that since we have been observing them, that is to say for about fifty years, we have not seen the *Coryanthes macrantha* or the *Catasetidae* perfect their trap; that is all we can affirm, and it is truly insufficient. Have we even attempted the most elementary experiments, and do we know what the successive generations of our amazing soaking orchid might do in a century's time if placed in a different milieu among strange insects? What is more, the names we give to the orders, species, and varieties end up fooling us, and so we create imaginary types we believe to be fixed, when in fact they are probably only the representatives of one and the same flower, which continues to modify its organs slowly in accordance with slow circumstances.

Flowers preceded insects upon this earth; when the latter appeared, the flower had therefore to adapt an entirely new mechanism to the habits of these unexpected collaborators. This fact alone, geologically indisputable, amid all that we do not know, is enough to establish evolution, and does not this rather vague word mean, in the final analysis, adaptation, modification, intelligent progress?

Moreover, it would be easy, without resorting to that prehistoric event, to group together a large number of facts showing that the faculty of adaptation and intelligent progress is not the exclusive domain of the human race. Without going back over the detailed chapters I have devoted to this subject in *The Life of the Bee*, I will simply recall two or three topical details which are mentioned there. Bees, for instance, invented the hive. In a wild and primitive state and in their country of origin, they work in the open air. It is the uncertainty, the inclemency of our northern seasons that inspired them to seek shelter in the hollows of rocks or trees. This stroke of genius restored the work of honey-gathering and care of the brood nest to the thousands of workers formerly immobilized around the honeycomb to maintain the necessary heat. It is not uncommon, especially in the Midi, during exceptionally mild summers, for them to revert to the tropical habits of their ancestors.*

---

*I had just written these lines, when M. E-L. Bouvier gave a paper at the Academy of Sciences (Record of 7 May 1906) on the subject of two open-air nestings verified in Paris, one in a *Sophora japonica*, the other in a horse chestnut tree. The latter, balanced on a small branch comprising two forks quite close to each other, was the more remarkable, on account of the evident and intelligent adaptation to particularly difficult circumstances.

The bees [I quote the summary of M. de Parville in the science section of *Débats*, 31 May 1906] constructed supporting pillars and had recourse to truly remarkable protective devices. They ended up transforming the double fork of the chestnut tree into a solid ceiling. Undoubtedly an ingenious human being would have done less well.

To protect themselves from the rain, they had installed enclosures and wadding, and blinds against the sun. One can only get an idea of the perfect industry of the bees by taking a close look at the architecture of the two nestings which are now in the museum.

Another fact: transported to Australia or California, our black bee completely changes its habits. Having grown certain of endless summer and of flowers never failing to appear, it lives, from the second or third year onward, content to gather from day to day only the honey and pollen necessary for daily consumption, and its recent and reasoned observation supplanting its hereditary experience, it makes no further provisions. In connection with this, Büchner mentions a trait that also proves adaptation to circumstances, not slow, ancient, unconscious, and inevitable but immediate and intelligent: on Barbados, amid refineries where they find sugar in abundance throughout the year, the bees wholly cease their visits to flowers.

Finally, let us recall the amusing contradiction the bees gave to two learned English entomologists: Kirby and Spence. "Show us," they said, "a single case where pressure of circumstances has inspired them to substitute clay or mortar for wax or propolis, and we will admit they are capable of reasoning."

Barely had they expressed this rather arbitrary wish than another naturalist, Andrew Knight, having smeared the bark of certain trees with a kind of cement made from wax and turpentine, observed that his bees completely gave up gathering propolis and used only the new and unknown substance, which they found ready-made and in abundance in the vicinity of their dwelling place. What is more, in the practice of apiculture, when there is a dearth of pollen, it is enough to supply the bees with a few pinches of flour for them to understand immediately that this can offer them the same services and be utilized in the same way as the dust of the anthers, although its taste, smell, and color are totally different.

What I have just recalled concerning bees might, I think, mutatis mutandis, be confirmed in the kingdom of flowers. It would probably be enough for the admirable evolutionary effort of the numerous varieties of sage, for example, to be subjected to some experiments and studied more methodically than I, a mere layman, am capable of. Meanwhile, among quite a few other indications that would be easy to collate, an unusual study of cereals by Babinet tells us that certain plants, when transported far from their habitual climate, observe the new circumstances and utilize them, exactly like the bees. Thus, in the hottest regions of Asia, Africa, and America, where the winter does not kill it annually, our wheat becomes again what it originally must have been: a plant as perennial as grass. It remains green, reproduces at root and no longer bears ears or grains. When, therefore, from its tropical and original home, it came to acclimatize itself to our icy lands, it must have needed to transform its habits and invent a new method of multiplication. Babinet puts it well: "The organism of the plant, by an inconceivable miracle, seemed to foresee the need of passing through the grain stage, in order not to perish completely during the harsh season."

XXIV

In any case, to put paid to the objection I spoke of earlier and which has caused this long digression, it would be enough to establish one single act of intelligent progress outside mankind. But apart from the pleasure we take in refuting an overly vain and outdated argument, this question of the personal intelligence of flowers, insects, or birds is basically of little importance! Even if we say, concerning the orchid and the bee alike, that it is nature and not the plant or the insect that calculates, combines, decorates, invents, and reasons, what interest can this distinction hold for us? A far greater question and one worthier of our full attention dominates these details. It is a matter of grasping the character, the quality, the habits, and perhaps the object of the general intelligence whence emanate all intelligent acts performed on this earth. It is from this point of view that the study of creatures—ants and bees, among others—in which are manifested most clearly, outside the human form, the processes and the ideal of that genius, is one of the most curious that we may undertake. It seems, after all we have just observed, that these tendencies, these intellectual methods are at least as complex, as advanced, as striking in the orchids as in the social *Hymenopterae*. Let us add that many of the intentions and a part of the logic of these swarming insects, so difficult to observe, still elude us, whereas we may grasp with ease all the silent motives, all the wise and stable arguments of the peaceful flower.

XXV

Now what do we observe, in catching nature at work? General intelligence or universal genius (the name matters little) in the world of flowers? Plenty of things, and to speak only in passing, for the subject would lend itself to a lengthy study, we ascertain right from the start that the flower's idea of beauty, of lightheartedness, its methods of seduction and its aesthetic tastes are very close to our own. But it would doubtless be more accurate to affirm that its and ours are in keeping. It is, in fact, highly uncertain that we have ever invented a beauty entirely our own. All our architectural and musical motifs, all our harmonies of color and light, etc., are borrowed directly from nature. Without evoking the sea, the mountains, the skies, the night, the dusk, what might one not say, for instance, of the beauty of trees? I speak not only of the tree as it makes up the forest, where it forms one of the great powers of the earth, perhaps the principal source of our instincts, of our sense of the universe, but of the tree in itself, of the solitary tree, whose green old age is laden with a thousand seasons. Among those impressions which, without our knowing it, form the limpid hollow and perhaps the deepest calm and happiness of our entire existence, who among us does not guard the memory of a few beautiful trees? When we have passed midlife, when we reach the end of the awestruck phase, when we have all but exhausted the spectacles that may be offered by art, genius, and the wealth of centuries and men, after having experienced and compared so many things, we return to very simple memories. Two or three innocent, unchanging, and refreshing images stand out on the purified horizon, images that we would like to carry into the final sleep, if it be true that an image may cross the threshold that separates our two worlds. For my part, I can imagine no paradise, nor life beyond the grave, however splendid it may be, in which one would fail to find in place some magnificent beech of the Sainte-Baume Mountains, some cypress, or some umbrella pine of Florence or of a humble hermitage close to my house, any one of which offers the passerby a model of all the great movements of necessary resistance, of quiet courage, of soaring, of gravity, of silent victory, and of perseverance.

XXVI

But I digress too much; I simply intended to note, concerning the flower, that nature, when it wishes to be beautiful, to please, to delight, and to show its joy, rather does as we would do were we endowed with its treasures. I know, speaking thus, that I speak somewhat like the bishop who admired the fact that providence always made great rivers flow close to great cities, but it is difficult to envisage these things from a nonhuman point of view. Let us, then, from this point of view, consider that we would know very few signs or expressions of happiness if we did not know the flower. In order to judge well its power of joy and beauty, we must live in that part of a country where it reigns unchallenged, like the corner of Provence, between the Stagne and the Loup, where I am writing these lines. Here, truly, the flower is the sole sovereign of the valleys and hills. The peasants here have lost the habit of growing wheat, as though they no longer had to provide other than for the needs of subtler humankind that would live on sweet scents and ambrosia. The fields form one great bouquet that renews itself endlessly, and the scents that follow one another seem to do their rounds throughout the sky-blue year. Anemones, gillyflowers, mimosas, violets, pinks, narcissi, hyacinths, daffodils, mignonettes, jasmines, and tuberoses invade the days and nights, the months of winter, summer, spring, and autumn. But the finest moment belongs to the May roses. Then, as far as the eye can see, from the hillsides to the hollow of the plains, between embankments of vines and olive trees, they flow on all sides like a river of petals whence emerge houses and trees, a river of the color we give to youth, health, and joy. The aroma that spreads across the sky, at once warm and fresh, but especially roomy, emanates, one would think, directly from the sources of bliss. The roadways and pathways are carved in the pulp of the flower, in the very substance of paradises. It seems that, for the first time in our life, we have a satisfying vision of happiness.

XXVII

Still from our human point of view and preserving the necessary illusion, let us add to our first remark another one a little more elaborate, a little less hazardous, and perhaps great with consequence, namely, that the genius of the earth, which is probably that of the entire world, acts in the vital struggle exactly as a man would do. It employs the same methods, the same logic. It reaches its goals by the means we would use, it feels its way, it hesitates, it goes in for trial and error, it adds, it eliminates, it recognizes and corrects its mistakes, as we would do in its place. It exerts itself, it invents laboriously and little by little, in the manner of the artisans and engineers of our workshops. It struggles, as we would, with the weighty, huge, and obscure mass of its being. It knows no more than we whither it goes; it seeks and finds itself gradually. It has an often confused ideal, but one in which we distinguish nonetheless a host of broad outlines that rise toward a more passionate, complex, sinewy, and spiritual form of life. Materially, it disposes of infinite resources, it knows the secret of prodigious forces of which we are unaware, but intellectually, it seems strictly to occupy our sphere, and we do not claim, thus far, that it exceeds its limits. And if it does not try to derive anything from beyond that sphere, does it not mean there is nothing outside it? Does it not mean that the methods of the human mind are the only ones possible, that man has not got it wrong, that he is neither an exception nor a monster, but the being through whom pass and in whom are manifested most intensely the great will, the great desire of the universe?

XXVIII

The benchmarks of our knowledge emerge slowly, sparingly. Perhaps Plato's famous image, the cave on whose walls unaccountable shadows are reflected, no longer suffices, but if we wanted to substitute for it a new and more exact figure, it would hardly be more consoling. Imagine that cave grown larger. No ray of light would ever enter it. Excepting daylight and fire, it would be carefully furnished with everything that comprises our civilization, and men would find themselves prisoners there from birth. They would not miss the light, having never seen it; they would not be blind, and their eyes would not be dead, but having nothing to look at, those eyes would probably become the most sensitive organ of touch.

In order to recognize ourselves in their movements, let us picture these unfortunates in their shadows, amid the multitude of unknown things that surround them. What strange mistakes, incredible deviations, unexpected interpretations! But how touching and often ingenious would seem the course of action they would take regarding things that had not been created for use at night! How many times would they have guessed correctly, and how great would be their astonishment, if suddenly, in the light of day, they were to discover the true nature and intended purpose of tools and devices that they would have tried their best to adapt to the uncertainties of the darkness?

Yet in comparison with our own, their position seems simple and easy. The mystery in which they creep about is limited. They are only deprived of one sense, whereas it is impossible to estimate the number of those in which we are lacking. The cause of their errors is unique, but we cannot keep count of our own.

Since we live in a cave of this kind, is it not interesting to assert that the power that has placed us there acts often and on some important points just as we ourselves act? There are gleams of light in our underground cavern that show us we have not been mistaken in the use of all the objects to be found therein, and some of these gleams are brought to us there by insects and flowers.

XXIX

We have taken for a long time a rather foolish pride in believing ourselves to be miraculous beings, unique and wonderfully open to chance, probably fallen from another world, without clear ties to the rest of life and, in any case, endowed with an unusual, incomparable, awful ability. It is far preferable to be nowhere near so prodigious, for we have learned that prodigies soon vanish in the normal evolution of nature. It is much more consoling to observe that we follow the same path as the soul of this great world, that we have the same ideas, the same hopes, the same trials, and—were it not for our specific dream of justice and pity—almost the same feelings. It is much more calming to assure ourselves that, to better our lot, to utilize the forces, the opportunities, the laws of matter, we employ methods exactly the same as those that the soul uses to illuminate and order its unruly and unconscious areas; that there are no other methods, that we are in the midst of truth, that we are in our rightful place and at home in this universe molded by unknown substances, whose thought is not impenetrable and hostile but analogous or apposite to our own.

If nature knew everything, if it were never mistaken, if everywhere, in all its enterprises, it showed itself from the outset to be perfect and infallible, if it revealed intelligence in every respect immeasurably superior to our own, then there would be reason to fear and to lose courage. We would feel like the victims and prey of an extraneous power, one which we would have no hope of knowing or measuring. It is far preferable to convince ourselves that this power, at the very least from an intellectual standpoint, is closely related to our own. Our mind draws from the same reservoirs as does that of nature. We are of the same world, we are almost among equals. We no longer mix with inaccessible gods, but with veiled and fraternal wills, which are for us to discover unexpectedly and to redirect.

XXX

It would not, I imagine, be very bold to maintain that we do not have beings who are more or less intelligent, but a scattered, general intelligence, a kind of universal fluid that penetrates diversely the organisms it encounters, depending on whether they are good or bad conductors of consciousness. Mankind would represent, until now, upon this earth, the realm of life that offered the least resistance to this fluid that the religions call divine. Our nerves would be the wires along which this subtler electricity would spread. The circumvolutions of our brain would form in some way the electric coil in which the force of the current would multiply, but this current would be of no other nature, would come from no other source than that which passes through the stone, the star, the flower, or the animal.

But these are mysteries that it is somewhat idle to question, seeing that we do not yet possess an organ that can obtain their response. Let us be content in having observed certain manifestations of that intelligence outside ourselves. Everything we observe in ourselves is rightly suspect; we are at once judge and plaintiff, and we have too much interest in peopling our world with magnificent illusions and hopes. But let the slightest external indication be dear and precious to us. Those that the flowers have just offered us are probably quite minor compared with those the mountain, the sea, and the stars might tell us, if we were to come upon the secrets of their life. They allow us nonetheless to presume with more self-confidence that the spirit animating all things or emanating from them is of the same essence as that one animating our bodies. If this spirit resembles us, if we thus resemble it, if all that is found in it is also found in ourselves, if it employs our methods, if it has our habits, our preoccupations, our tendencies, our desires for improvement, is it illogical to hope for all we hope for, instinctively, irresistibly, since almost certainly it hopes for the same too? Is it likely, when we find scattered through life such a great sum of intelligence, that this life should make no work of intelligence, that is to say, should not pursue a goal of happiness, of perfection, of victory over that which we call evil, death, darkness, nothingness, that which is probably only the shadow of its face or its own state of sleep?

Scents

After having spoken long enough about the intelligence of flowers, it will seem natural for us to say a word about their soul, which is their scent. Unfortunately here, as with the soul of man, the scent of another sphere in which reason bathes, we immediately touch upon the unknowable. We know virtually nothing of the intention of that magnificently invisible zone of festive air that the corollas spread around themselves. It is, in fact, highly doubtful that it serves principally to attract insects. First of all, many flowers, among them the most fragrant, do not allow cross-fertilization, so that the visit of the bee or the butterfly is a matter of indifference or is unwelcome to them. Then, what draw the insects are strictly the pollen and the nectar, which generally have no discernible scent. So we see them neglect the most deliciously scented flowers, like the rose and the pink, in order to throng those of the maple or the hazel, whose aroma is virtually nonexistent.

Let us admit, then, that we still do not know in what way scents are useful to the flower, just as we do not know why we detect them. Smell is actually the most unexplained of our senses. It is evident that sight, hearing, touch, and taste are indispensable to our animal life. Only a lengthy education teaches us to take pleasure impartially in form, color, and sound. Moreover, our sense of smell also performs important menial functions. It is the guardian of the air we breathe; it is the hygienist and the chemist who watch carefully over the quality of the proffered foodstuffs, every unpleasant emanation revealing the presence of suspicious or dangerous germs. But alongside this practical mission, there is another that seemingly answers to nothing. Scents are in all respects useless to our physical life. Too violent, too permanent, they can even become hostile to it. We possess nonetheless a faculty that takes great pleasure in them and brings us good tidings with as much enthusiasm and conviction as if it were a matter of discovering a fruit or a delicious beverage. This uselessness deserves our attention. It must hide a beautiful secret. Here is the sole occurrence of nature that procures us a gratuitous pleasure, a satisfaction that does not adorn a trap of necessity. Smell is the unique sense of luxury granted to us. But it seems almost foreign to our body, tied very loosely to our organism. Is it a device that develops or decays, a faculty that is falling asleep or awakening? Everything leads us to believe that it evolves along with our civilization. The ancients barely bothered with anything other than the most brutal, heavy, solid smells, so to speak—musk, benzoin, myrrh, frankincense, etc.—and the scent of flowers is quite rarely mentioned in Greek and Latin poetry or in Hebrew literature.

Yet do we see our peasant folk today, even at their greatest ease, think of sniffing a violet or a rose? On the contrary, is it not the very first gesture of the big-city inhabitant who comes across a flower? There is sufficient reason therefore to accept smell as the last born of our senses, the only one perhaps that may not be "on the road to regression," as the biologists so thoughtfully put it. It is a reason for us to attach ourselves to it, question it, and cultivate its potential. Who may tell what surprises it would have in store for us, if it were to equal, for instance, the perfection of the eye, as it does in the dog, who lives as much by the nose as by the eyes?

We have there an unexplored world. To think better of this mysterious sense that, at first, seems almost foreign to our organism, is perhaps to think of it as the most intimately penetrative one. Are we not, above all, aerial beings? Is not air the most absolutely and most immediately indispensable element to us, and is not smell truly the only sense that perceives some parts of it? Scents, jewels of the air we need to breathe, do not embellish it without reason. It would not be surprising if this misunderstood luxury answers to something deeply profound and essential and rather, as we have just seen, to something that does not yet exist than to something that no longer does. It is highly likely that this sense, the only one turned toward the future, may already grasp the most striking manifestations of a form or of a happy or beneficial state of matter that holds plenty of surprises in store for us.

Meanwhile, it remains subject to the most violent, least subtle perceptions. It barely suspects, in making use of the imagination, the profound and harmonious emanations that clearly envelop the great spectacles of atmosphere and light. As we are on the verge of grasping those of rain or of dusk, why should we not come to discern and to determine the scent of snow, ice, morning dew, the first fruits of dawn, the twinkling of stars? Everything in space must have its scent, still inconceivable, even a moonbeam, a trickle of water, a drifting cloud, a smiling sky.

Chance, or rather choice of life, has led me back these days to the places where nearly all the scents of Europe are born and prepared. In fact, as we all know, it is on the luminous strip of land stretching from Cannes to Nice that the last hills and last valleys of authentic living flowers keep up a heroic struggle with the foul chemical odors of Germany, which are exactly to natural scents what painted forests and plains in an art gallery are to the forests and plains of the true countryside.

The work of the peasant there is ordered by a kind of uniquely floral calendar, in which, in May and July, two adorable queens reign supreme: the rose and the jasmine. From January to December, around these two sovereigns of the year, one the color of dawn, the other clothed in white stars, parade the innumerable and hasty violets, the tumultuous daffodils, the naive narcissi that fill the eye with wonder, the giant mimosas, the mignonettes, the pinks laden with precious spices, the imperious geranium, the tyrannically virginal orange blossom, the lavender, the Spanish broom, the all too powerful tuberose, and the cassia, which is a type of acacia and bears a flower similar to an orange-hued caterpillar.

It is rather disconcerting at first to see the large, dull-witted, and uncouth rustics, whom pure necessity everywhere distracts from the smiling side of life, taking flowers so seriously, handling so carefully these fragile ornaments of the earth, accomplishing a bee's or a princess's task and bending beneath the yoke of the violets or the daffodils. But the most striking impression is that of certain evenings or mornings in the season of the rose or the jasmine. We might believe the atmosphere of the earth has undergone a sudden change, has given way to that of an infinitely happy planet, where scent is no longer, as down here, fleeting, vague, and precarious, but stable, expansive, full, permanent, generous, normal, inalienable.

In speaking of Grasse and its surroundings, we have drawn more than once—at least I suppose so—the picture of that almost fairy-tale industry occupying an entire hardworking town, stuck on the side of a mountain like a sun-drenched hive. We must have spoken of the magnificent cartloads of pink roses unloaded on the doorsteps of the smoking factories, the vast halls where the sorters are literally swimming in the waves of petals, the less bulky but more precious arrival of the violets, the tuberoses, the cassias, the jasmines, in broad baskets that the peasant women carry nobly on their heads. We must have described the various procedures by which we extract from the flowers, according to their features, the marvelous secrets of their hearts, so as to fix them in crystal. We know that some roses, for example, are extremely obliging and good-willed and yield their aroma straightforwardly. We pile them up in huge boilers, as tall as those of our rail locomotives, into which steam passes. Little by little their essential oil, more costly than a jelly of pearls, seeps drop by drop into a narrow glass tube like a goose feather, at the bottom of an alembic resembling some kind of monster that would with difficulty give birth to an amber tear.

But the majority of flowers let their soul be imprisoned less easily. I shall not speak here of all the infinitely varied tortures that we inflict on them to force them eventually to abandon the treasure they hide desperately at the bottom of their corollas. To give an idea of the trick of the executioner and of the obstinacy of certain victims, it will suffice to recall the agony of the cold press that the daffodil, the mignonette, the tuberose, and the jasmine undergo before breaking the silence. Let us note in passing that the scent of the jasmine is the only inimitable one, the only one unobtainable by the skilful blending of other scents.

So we spread out a bed of grease thick as two fingers on large glass plates, and the whole is abundantly covered with flowers. Following what unctuous maneuvers and promises does the grease obtain the irrevocable confidences? Whatever the case may be, the poor, all too trusting flowers soon have nothing left to lose. Each morning they are picked up, thrown into the garbage, and a new swathe of ingénues replaces them on the insidious bed. In turn they succumb, suffer the same fate, and then many others follow them. It is only after three months, that is to say, after having devoured ninety generations of flowers that the greedy and specious grease, sated with surrenders and with embalmed confessions, refuses to strip any further victims.

As for the violet, it resists the entreaties of the cold grease; we must add to it the torment of fire. We heat the lard in a pot. Following this barbaric treatment, the humble and pleasant flower of the springtime byways gradually loses the strength that kept its secret. It yields, it gives itself up, and its liquid executioner, before being sated, absorbs four times its weight in petals, which means that the shameful torture continues throughout the season of the violets blooming beneath the olive trees.

But the drama is not over. Now it is a matter, whether it is warm or cold, of making this miserly grease cough up the absorbed treasure, which it intends to hold on to with all its crude and evasive energies. We succeed in this with great difficulty. The grease has base passions that ruin it. It is soaked in alcohol, it is intoxicated, and it ends up by letting go. Now it is the alcohol that possesses the mystery. No sooner has the alcohol seized the mystery than it too intends to share it with no one else, to keep it for itself alone. In turn we attack it, we reduce it, we evaporate it, we condense it, and the liquid pearl, after so many adventures, pure, essential, inexhaustible, and almost imperishable, is finally gathered in a crystal phial.

I will not list the chemical extraction processes: by petroleum gases, by carbon sulfur, etc. The great perfumers of Grasse, loyal to tradition, are repelled by those artificial and almost dishonest methods, which yield only pungent scents and bruise the flower's soul.

On the Publication History
of Maeterlinck's Botanical Essays

In 1904 Maeterlinck broached a botanical theme in *Le Double Jardin* (The Double Garden), a collection of sixteen essays on various subjects published in Paris by Eugène Fasquelle. It contained four essays related to botany: "Les Sources du printemps" (The Sources of Spring), "Fleurs des champs" (Flowers of the Fields), "Crysanthèmes" (Crysanthemums), and "Fleurs démodées" (Old-fashioned Flowers). Incidentally, this collection also contained one of Maeterlinck's most celebrated essays, "Sur la mort d'un petit chien" (On the Death of a Small Dog).

L'Intelligence des fleurs was first published in Paris by Eugène Fasquelle in 1907. Including "Les Parfums" (Scents), it contained eleven essays, again on various subjects. The eponymous essay first appeared in English in *The Intelligence of the Flowers* translated by Alexander Teixeira de Mattos in numbers 679, 681 and 682 of *Harper's Magazine* in 1906–07. Also in 1907 this translation was published as a book in New York by Dodd, Mead and Co., with a twelfth essay added, decoration by William Edgar Fisher, and illustrations from photographs by Alvin Langdon Coburn. Later that same year Dodd, Mead and Co. added another edition, entitled *The Measure of the Hours*, changing the order of the essays and naming the volume after the third essay, whose subject is time.

L'Intelligence des fleurs was republished in 1923 jointly by Robert Sand in Brussels, Editions du Dauphin in Antwerp, and G. Crès in Paris. This edition included woodcuts by J-M. Canneel and a frontispiece etched by Louise Danse. Further editions appeared in 1939 (Fasquelle), reprinted in facsimile in 1977 by Editions d'Aujourd'hui (Paris); in 1946 in Lausanne (Editions du Grand-Chêne); in 1954, in a volume entitled *Insectes et fleurs* (Insects and Flowers, Paris: Gallimard/N.R.F.) along with the corresponding essays on the bee, the ant, and the termite plus a 1934 essay on the pigeon; and in 1955 (Paris: Editions du Reflet) with illustrations by Tavy Notton. In 1997, the essay (again along with those on the bee, the ant, and the termite) appeared in a volume entitled *La Vie de la nature* (The Life of Nature, Brussels: Editions Complexe) with a preface by Jacques Lacarrière and an afterword by Paul Gorceix.

Republications of the English translation of *L'Intelligence des fleurs* appeared in 1911, as *Life and Flowers* (London: George Allen), a British edition of the original American translation, also with a twelfth essay added; and in 2001 (Honolulu: University Press of the Pacific), a reprint of the 1907 translation of the eponymous essay.

The present translation seeks to render Maeterlinck's essay more accessible to the contemporary reader and to offer a critical context lacking in earlier editions.

Select Bibliography

ON MAETERLINCK

Chauvin, Rémy. "Le Point de vue d'un biologiste." *Europe*, 399–400 (July–August) 1962. Special issue on Maeterlinck.

De Corte, Marcel. "Le Problème de l'intelligence chez les insectes et chez les fleurs." In *Brochures-programmes de l'Institut National de Radiodiffusion* [Brussels] (Hommage à Maurice Maeterlinck, 1862–1949), 1949.

Gorceix, Paul. "Postface." In Maurice Maeterlinck, *La Vie de la nature*. Brussels: Editions Complexe, 1997.

Halls, W. D. *Maurice Maeterlinck: A Study of His Life and Thought*. Oxford: Clarendon Press, 1960.

Knapp, Bettina. *Maurice Maeterlinck*. Boston: Twayne, 1975.

Konrad, Linn B. *Modern Drama as Crisis: The Case of Maurice Maeterlinck*. Frankfurt: Peter Lang, 1986.

Lecat, Maurice. *Le Maeterlinckianisme*. Brussels: Castaigne, 1937.

Mahony, Patrick. *Maurice Maeterlinck: Mystic and Dramatist*. Washington, D.C.: Institute for the Study of Man, 1984.

McGuinness, Patrick. *Maurice Maeterlinck and the Making of Modern Theatre*. Oxford: Oxford University Press, 2000.

Pasquier, Alex. *Maurice Maeterlinck*. Brussels: La Renaissance du Livre, 1950.

Présence/Absence de Maurice Maeterlinck. Actes du Colloque de Cerisy-la-Salle, 2000. Brussels: Archives et Musée de la Littérature, 2002.

Rostand, Jean. "Hommage d'un naturaliste à Maurice Maeterlinck." In *Maurice Maeterlinck 1862–1962*, ed. Joseph Hanse and Robert Vivier. Brussels: La Renaissance du Livre, 1962.

Rykner, Arnaud. *Maurice Maeterlinck*. Rome: Memini, 1998.

Stowe, Richard S. "Maurice Maeterlinck." In *European Writers: The Twentieth Century*, vol. 8, ed. George Stade. New York: Charles Scribner's Sons, 1989.

Taylor, Una. *Maurice Maeterlinck: A Critical Study*. London: Martin Secker, 1914.

Thomas, Edward. *Maurice Maeterlinck*. New York: Dodd, Mead and Co., 1911.

Timmermans, B. *L'Evolution de Maeterlinck*. Brussels: La Belgique Artistique et Littéraire, 1912.

Turquet-Milnes, G. *Some Modern Belgian Writers: A Critical Study*. Freeport, N.Y.: Books for Libraries Press, 1968 [1917].

Vallet, Jacques. "Pour embrasser les vaches." *Textyles* 1 (September) 1985.

Vivier, Robert. "Deux aspects de Maeterlinck." In *Le Centenaire de Maurice Maeterlinck*. Brussels: Palais des Académies, 1964.

ON SCIENCE AND THE HUMANITIES

Barthes, Roland. "Science versus Literature." In *Structuralism: A Reader*, ed. Michael Lane. London: Jonathan Cape, 1970 [1967].

Beer, Gillian. *Darwin's Plots: Evolutionary Narrative in Darwin, George Eliot and Nineteenth-Century Fiction*. London: Routledge and Kegan Paul, 1985.

————. *Open Fields: Science in Cultural Encounter*. Oxford: Clarendon Press, 1996.

Chapple, J. A. V. *Science and Literature in the Nineteenth Century*. Basingstoke: Macmillan, 1986.

Christie, John, and Sally Shuttleworth. *Nature Transfigured: Science and Literature, 1700–1900*. Manchester: Manchester University Press, 1989.

Deery, June. *Aldous Huxley and the Mysticism of Science*. New York: St. Martin's Press, 1996.

Foucault, Michel. *The Order of Things: An Archaeology of the Human Sciences*. New York: Pantheon Books, 1971.

Hayles, N. Katherine. *The Cosmic Web: Scientific Field Models and Literary Strategies in the Twentieth Century*. Ithaca: Cornell University Press, 1990.

————. *Chaos Bound: Orderly Disorder in Contemporary Literature and Science*. Ithaca: Cornell University Press, 1990.

Hyman, Stanley Edgar. *The Tangled Bank: Darwin, Marx, Frazer and Freud as Imaginative Writers*. New York: Atheneum, 1962.

Levine, George. *Darwin and the Novelists: Patterns of Science in Victorian Fiction*. Cambridge, Mass.: Harvard University Press, 1988.

Locke, David. *Science as Writing*. New Haven: Yale University Press, 1992.

Merleau-Ponty, Maurice. *Nature: Course Notes from the Collège de France*, trans. Robert Vallier. Evanston, Ill.: Northwestern University Press, 2003.

Paradis, James, and Thomas Postlewait, eds. *Victorian Science and Victorian Values: Literary Perspectives*. New Brunswick: Rutgers University Press, 1985.

Serres, Michel. *Hermes: Literature, Science, Philosophy*, ed. Josué V. Harari and David F. Bell. Baltimore: Johns Hopkins University Press, 1982.

Snow, C. P. *The Two Cultures: and A Second Look*. Cambridge: Cambridge University Press, 1964.

Sorell, Tom. *Scientism: Philosophy and the Infatuation with Science.* New York: Routledge, 1991.

Thiher, Allen. *Fiction Rivals Science: The French Novel from Balzac to Proust.* Columbia: University of Missouri Press, 2001.

Turner, Frank Miller. *Between Science and Religion: The Reaction to Scientific Naturalism in Late Victorian England.* New Haven: Yale University Press, 1974.

On Humanistic Botany

Attenborough, David. *The Private Life of Plants: A Natural History of Plant Behavior.* Princeton: Princeton University Press, 1995.

Buhner, Stephen Harrod. *The Lost Language of Plants: The Ecological Importance of Plant Medicines for Life on Earth.* White River Junction, Vt.: Chelsea Green, 2002.

Farr, Judith. *The Gardens of Emily Dickinson.* Cambridge, Mass.: Harvard University Press, 2004.

Iseley, Duane. *One Hundred and One Botanists.* Ames: Iowa State University, 1994.

Jordan, Michael. *The Green Mantle: An Investigation into Our Lost Knowledge of Plants.* London: Cassell, 2001.

Kincaid, Jamaica. *My Favorite Plant: Writers and Gardeners on the Plants They Love.* New York: Farrar, Straus and Giroux, 1998.

———.. *My Own Garden.* New York: Farrar, Straus and Giroux, 1999.

King, Amy M. *Bloom: The Botanical Vernacular in the English Novel.* New York: Oxford University Press, 2003.

Knapp, Sandra. *Plant Discoveries: A Botanist's Voyage through Plant Exploration.* New York: Firefly Books, 2003.

Mabey, Richard. *The Frampton Flora.* New York: Prentice Hall, 1986.

———.. *Flora Britannica.* London: Chatto and Windus, 1997.

Pollan, Michael. *The Botany of Desire: A Plant's-Eye View of the World*. New York: Random House, 2001.

Prusinkiewicz, Przemyslaw, and Aristid Lindenmayer. *The Algorithmic Beauty of Plants*. New York: Springer-Verlag, 1990.

Tompkins, Peter, and Christopher Bird. *The Secret Life of Plants*. New York: Harper and Row, 1973.